T0225095

Building Your Career as a Statistician

This book is intended for anyone who is considering a career in statistics or a related field, or those at any point in their career with sufficient work time remaining such that investing in additional learning could be beneficial. As such, the book would be suitable for anyone pursuing an MS or a PhD in statistics or those already working in statistics. The book focuses on the non-statistical aspects of being a statistician that are crucial for success. These factors include (1) productivity and prioritization, (2) innovation and creativity, (3) communication, (4) critical thinking and decisions under uncertainty, (5) influence and leadership, (6) working relationships, and (7) career planning and continued learning. Each of these chapters includes sections on foundational principles and a section on putting those principles into practice. Connections between these individual skills are emphasized such that the reader can appreciate how the skills build upon each other leading to a whole that is greater than the sum of its parts. By including the individual perspectives from other experts on the fundamental principles and their application, readers will have a well-rounded view of how to build upon and fully leverage their technical skills in statistics. The primary audience for the book is large and diverse. It will be useful for self-study by virtually any statistician, but could also be used as a text in a graduate program that includes a course on careers and career development.

Key Features:

- Proven principles useful in and applicable to statistical and other settings
- Focused
- Concise
- Accessible to all levels, from graduate students to mid-later career statisticians

Building Your Career as a Statistician

A Practical Guide to Longevity, Happiness, and Accomplishment

Edited by

Craig Mallinckrodt

CRC Press
Taylor & Francis Group
Boca Raton London New York

CRC Press is an imprint of the
Taylor & Francis Group, an **informa** business
A CHAPMAN & HALL BOOK

Designed cover image: © Shutterstock, ID: 1328365757, Viktoria Kurpas

First edition published 2024
by CRC Press
6000 Broken Sound Parkway NW, Suite 300, Boca Raton, FL 33487-2742

and by CRC Press
4 Park Square, Milton Park, Abingdon, Oxon, OX14 4RN

CRC Press is an imprint of Taylor & Francis Group, LLC

ISBN: 9781032368771 (hbk)
ISBN: 9781032368795 (pbk)
ISBN: 9781003334286 (ebk)

DOI: 10.1201/9781003334286

Typeset in Minion
by Deanta Global Publishing Services, Chennai, India

To Donna and Marissa:

Your humor enlivens me.

Your kindness inspires me.

Your wisdom informs me.

Best of all, it's been a good feeling to know somebody loved me.

———————————————

Contents

SECTION III **Planning and Growing**

CHAPTER 9 ■ Career Planning and Continued Learning 185

SECTION IV **Other Perspectives**

List of Tables

List of Figures

Acknowledgments

I owe many thanks to Christy Chuang-Stein, Marc Buyse, and Geert Molenberghs, who each authored a chapter in this book. The diversity, perspective, and insight in their chapters are foundational to the topics covered throughout the book.

Thanks also to my long-time Lilly and Biogen colleague, Simon Cleall. I have had many wonderful chats with Simon over the years. He influenced my thinking on several important issues, including careers in statistics. I first heard the 'careers are like cars' analogy that is explained in Chapter 1 by Simon.

Preface

IN A JOB INTERVIEW, a hiring manager told a prospective employee: You have the perfect background for this job; the other 90% of what you need to know you can learn on the job. On the one hand, this is an intimidating statement, implying that all the hard work and learning that the prospective employee had done prior to the new job would count only a small amount toward success in the new role. But, on the other hand, what a great opportunity! The combination of the right background with the opportunity to learn and grow can be a great recipe for success.

The 10%/90% ratio of prior learning to new learning may be a bit of an exaggeration to make a point. Whether the exact ratio is 10%/90%, 50%/50%, 70%/30%, we need to continue to grow and learn to have long, happy, and productive careers. We have many opportunities for continued statistical learning. The intent of this book is to provide guidance on how to make the most of those skills.

A key premise of this book is that the success and happiness we achieve in our careers is attributable to both our technical acumen and a variety of non-statistical aspects of our jobs. To put this idea into perspective, Chapter 1 develops the analogy of a statistical career being like a car wherein statistical acumen is the engine that generates the raw power for our careers, with other features of the car (our careers) needed to translate that raw power into meaningful production.

Part I of the book continues with Chapters 2 through 5 that cover foundational principles and their application in increasing/improving our individual skills: Working with self. Topics covered include: productivity, prioritization, and work–life balance (Chapter 2), creativity and innovation (Chapter 3), communication (Chapter 4), and individual factors in critical thinking and making decisions under uncertainty (Chapter 5). Part II focuses on skills needed to work with others: Group factors in critical thinking and making decisions under uncertainty (Chapter 6), leadership (Chapter 7), and working relationships (Chapter 8). Part III covers continued learning (Chapter 9), and two chapters that summarize and integrate ideas from the previous chapters.

Most of these chapters conclude with a section on putting the principles into practice and a section on relevant examples. Part IV includes chapters contributed by statisticians that have had long, happy, and accomplished careers. Christy Chuang-Stein, Marc Buyse, and Geert Molenberghs share their perspectives on careers and career development. Their chapters broaden and reinforce the ideas from the earlier parts of the book. This diversity in perspectives is important because it allows us to see how the basic principles were applied by statisticians who built long, happy, and successful careers.

You will not find many specific answers to specific questions here because individual situations can be idiosyncratic. The intent of this book is to provide the basis from which you can find the answers that are right for you.

Editor

CRAIG MALLINCKRODT IS A Distinguished Biostatistician at Pentara Corporation. He is a Fellow of the American Statistical Association and he won the Royal Statistical Society's award for outstanding contribution to the pharmaceutical industry. He has extensive industry and academic working experience, holding technical positions from entry level to Vice President. Craig has led industry working groups and he has published extensively on a wide variety of clinical and methodological topics, most notably, missing data, estimands, and various aspects of clinical development optimization.

Contributors

Marc Buyse
IDDI and University of Hasselt
Hasselt, Belgium

Christy Chuang-Stein
Chuang-Stein Consulting, LLC
Kalamazoo, USA

Geert Molenberghs
University of Hasselt and KU Leuven
Hasselt, Belgium

I

Working with Self

CHAPTER 1

Introduction

ABSTRACT

This chapter contrasts two stories, one of Abraham Wald who was able to figure out a tough problem, convince others of his findings, and implement them successfully. The other story was of Ignaz Semmelweis, who also figured out a tough problem but was not able to convince others and successfully implement the findings. The point of the stories is that being right isn't enough. This chapter uses the analogy of a car and its many features as a model for careers in statistics. Technical acumen is the engine that powers our careers. But just as being right isn't enough to ensure success, just as an engine is not the only important component of a car, there is more to being a statistician than technical acumen. Chapter 1 introduces the other features that statisticians must possess at least to some degree to have a long, happy, and accomplished career. The chapter also provides a brief background about the author and his career, along with some initial perspectives on success.

DOI: 10.1201/9781003334286-2

1.1 BULLET HOLES AND LISTERINE

In the book *How Not to Be Wrong* (Ellenberg, 2014), the author Jordan Ellenberg summarizes the following story about thinking and quantitative analysis. During the second world war, the US Air Force was using data analysis – statistics and modeling – to inform efficient armoring of their planes. The exterior of the planes was vulnerable to anti-aircraft fire. Protective armor could not be added everywhere because the increased weight would hinder flight performance and fuel economy (flight range).

A single bullet strike was usually not catastrophic to flight. In fact, in many areas, multiple strikes were not catastrophic. Therefore, the bullet holes in planes that returned to the base were used to ascertain the most frequently struck areas, and these were the areas to be armored. A representation of the bullet strikes is provided in Figure 1.1. If the story ended here, it would be about how modeling and statistics contributed to a life-saving solution. But the story continues because the initial conclusion was wrong.

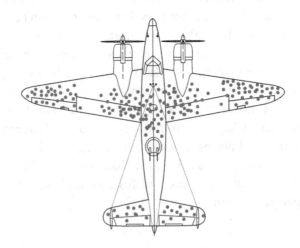

FIGURE 1.1 Diagram representing the pattern of bullet holes in returning airplanes.

Abraham Wald, the famous mathematician/statistician, concluded that armor should go where the bullet holes were not (or were infrequent)! Wald was using the same data as others, but he realized that these data were biased. In the returning planes, bullet holes were least frequent in the engine compartment because planes with damaged engines were less likely to return to base and have their bullet holes counted. Wald convinced others of his conclusion and armor was added to the engine compartment.

In another story of great technical acumen, the outcome was not so good. Ignaz Semmelweis (1818–1865) was a physician and scientist. In the mid-19th century, puerperal fever was common in hospitals, and it was often fatal. In fact, the fatality rate of this so-called childbed fever was threefold greater in hospitals than in midwives' wards. Semmelweis discovered that the incidence of puerperal fever was reduced by the use of hand disinfection in obstetrical clinics. Data for this observation are summarized in Figure 1.2.

Despite various publications that also showed handwashing reduced mortality, Semmelweis's observations conflicted with established opinions and his ideas were rejected. Semmelweis could not explain why handwashing worked. Some doctors were offended at being seen as part of the problem rather than the

Table 1. Annual births, deaths, and mortality rates for all patients at the two clinics of the Vienna maternity hospital from 1841 to 1846.

| | First Clinic | | | Second Clinic | | |
	Births	Deaths	Rate	Births	Deaths	Rate
1841	3036	237	7.7	2442	86	3.5
1842	3287	518	15.8	2659	202	7.5
1843	3060	274	8.9	2739	164	5.9
1844	3157	260	8.2	2956	68	2.3
1845	3492	241	6.8	3241	66	2.03
1846	4010	459	11.4	3754	105	2.7
Total	20 042	1989		17 791	691	
Avg.			9.92			3.38

FIGURE 1.2 A table from the *Etiology, Concept, and Prophylaxis of Childbed Fever* – Ignaz Semmelweis, 1861.

solution to it. Semmelweis was mocked and ridiculed. In 1865, Semmelweis suffered a nervous breakdown and was committed to an asylum. In the asylum, he was beaten by the guards and died 14 days later from a gangrenous wound on his right hand that may have been caused by the beating.

Semmelweis's conclusions gained acceptance when Louis Pasteur confirmed the germ theory of disease, and Joseph Lister followed Pasteur's work and used hygienic methods to lower infection and mortality. Semmelweis was right, but he was unable to influence practice. He was right, but being right was not enough and today we have Listerine, not Semmelweisine.

These stories highlight several key points in how to build happy, long, and accomplished careers. We cannot be certain how Wald developed the key insights that allowed him to solve a problem that others got wrong. However, it was likely from a deep understanding of both statistics and the data – the subject matter of the investigation. Statistical knowledge was not sufficient, nor was knowledge of aviation. Solving the problem required knowledge of both. And being right was not enough. Wald had to convince others, he had to communicate, influence, and lead.

It is unclear why Semmelweis was not able to change medical practice. Perhaps he was too far ahead of his time. Without the requisite germ theory, his ideas lacked scientific footing. This may be an example of 'The Adjacent Possible' that is covered in Chapter 3 on creativity and innovation. The adjacent possible is an idea in which innovation moves in incremental steps rather than leaps and bounds. For Semmelweis' theory to be accepted, the medical community had to leap-frog past germ theory straight to disinfection. Maybe it was too far of a leap. Maybe there were other things Semmelweis could have done to garner more support. The record is not clear.

Nevertheless, Semmelweis' story highlights that technical acumen is not enough. As we say, it is necessary but not sufficient. In fact, some of the most unhappy statisticians I have known are those with strong technical ability but who felt they were unproductive, or unable to influence. Hence, an important focus of this book is on the non-statistical aspects of our work.

1.2 MY BACKGROUND

By my own measuring stick, I have had a long, happy, and accomplished career. But that doesn't mean it has always been smooth sailing. In fact, at times the disappointments and difficulties seemed to outnumber the successes. On the good side, I graduated with the highest distinction from my undergraduate program and had wonderful experiences in graduate school. I worked my way from a non-tenure track position in academia to an entry-level statistics position in the pharmaceutical industry, eventually to a Vice President role. I was elected an American Statistical Association (ASA) Fellow, won the Royal Statistical Society's Award for outstanding contribution to the pharmaceutical industry, helped develop drugs that have been used to treat millions of patients, and met and worked with many outstanding statisticians and physicians.

On the not-so-good side, I have a congenital vision impairment that resulted in me never being able to see what teachers were presenting in school, what was being projected on screens in meeting rooms, nor was I able to drive. However, this vision impairment was not the root cause of any of my failures, such as almost failing science in the seventh grade and chemistry in the tenth grade. Nor was the poor vision the reason why I went broke running a small business. After completing my PhD, I did not get the initial job I wanted. My first post-graduate job was a non-tenure-track academic position, and my position was terminated due to budget cuts. In a subsequent job search, I was given a typing test during an interview for a position requiring a PhD because the interviewer did not believe someone with my degree of vision impairment could use a computer.

I have thought a lot about what factors contributed to the good and the not so good in my career. I have read books with lots of advice helpful to statisticians. What I have learned can be summed up by paraphrasing the 18th-century geneticist Robert Bakewell: The difficulties lie not so much in knowing the principles, but rather in putting those principles into useful practice.

Throughout this book, I cite many ideas and principles that are useful in having a long, happy, and accomplished career in statistics. Some of these things I learned from others, some from my readings, and others from experience. I hope that sharing these ideas and my own career journey, along with advice from other expert statisticians, will help you put those principles into useful practice in your own journey.

1.3 CAR MODEL OF CAREERS

To understand how statistical and non-statistical factors contribute to our success, consider statistical careers to be like cars. I first encountered this analogy from a colleague, Simon Cleall.

- Statistical acumen is the engine that generates the raw power for careers in statistics, with other features needed to translate that raw power into meaningful production.

- The ability to work productively and to prioritize is engine efficiency – how much useful energy versus wasted energy and exhaust the engine generates. Being efficient at work provides the time needed for work–life balance.

- Communication skill is the transmission, how useful energy (knowledge, ideas, etc.) from the engine is transferred to move the car (team, audience) in the intended direction.

- Critical thinking and making decisions under uncertainty are the navigation, computer, and system monitors used to plan, navigate, and execute the route.

- Relationship skills are the sensors and driver-assisted systems that allow us to interact in traffic, to avoid obstacles, and to interact with others in the workplace.

- Influence and leadership are the suspension, steering, and entertainment systems to keep the vehicle (bus or van carrying multiple passengers) on the desired path and engaged, and to smooth out bumps along the way.

- Career development and continued learning are improvements, analogous to regular maintenance, model updates, and new models (Figure 1.3).

Like cars, we each can have different combinations of, or emphasis on, certain features. No one is expert in all features, especially at a single point in our careers. It is up to us to decide what combination of features we wish to develop, or which features are most important at a particular point in our career. Some of those features focus on working with self and our own skills. Other features focus more so on working with others.

FIGURE 1.3   An automotive assembly line. Careers are like cars, they are assembled from a collection of individual parts that work well together.

The feature we want to emphasize the most may change over time. Early in careers focus may be on productivity, whereas later in careers critical thinking and leadership become more important. If you do not work to develop a 'feature' your car is unlikely to have a good one; and, if you fail to maintain a feature, it may breakdown in the future.

Hence, an important aspect of having a long, happy, and accomplished career is continuing to work on our features, both to develop new or improved features and to maintain the features we already have. As we will see in the next chapter, maintenance is used here in a broad sense. We are talking about more than just life at work. If we do not maintain a proper work–life balance, what we do at work will provide little satisfaction.

1.4 PERSPECTIVE

Three general landmarks in career building are whether we are:

- motivated

- working smart

- working well with others

It is tempting to replace motivated with working hard; that is, the three questions to ask: Are you working hard, working smart, and working well with others. But that would be a mistake. We tend to focus too much on how hard and how long we work, which leads to us working too long and too hard, and focusing on activity rather than accomplishment. Again, using the car analogy, the optimum car is not one that works hard and long to get to the destination, but rather one that is efficient and dependable.

If we are motivated, we will work hard enough. If we are mindful that what we do outside of work is more important than what we do at work, we will keep the balance needed in our lives for happiness and longevity. For most statisticians, working harder is not the answer, it is the problem.

That is why much of this book has to do with working smart and working well with others. Nevertheless, before we get to those aspects in subsequent chapters, it is necessary to consider the foundational role of motivation. If we are not motivated and engaged in our work it is unlikely that we will have a long, happy, or productive career; it is unlikely we will have the energy and discipline needed to work smart and to work well with others. If we are not interested in and inspired by the work in our current job, or if the main demands of the job do not match our interests or skills, we may fare better in a different role.

Switching roles or seeking a new job is the topic beyond the scope of this chapter. However, it is worth noting a few general points. Changing jobs and roles has a cost that is exemplified by the flywheel effect described in the book *Good to Great* (Collins, 2001). The flywheel effect is when success comes not from a single event, action, program, innovation, or lucky break, but from consistent effort toward a goal. The metaphor Collins uses is that we push toward a goal as we would push a giant, heavy flywheel, turn after turn, building momentum until a point of breakthrough. Changing jobs or roles may cause a loss of momentum. Moreover, changing jobs or roles may result in losing out on key learnings and insights because a project was not followed through to completion. On the other hand, when one is not generating momentum, it may be time to consider alternatives.

Returning to the car analogy, a heavy-duty four-wheel-drive pick-up is not ideal for urban commuting and a small sedan is not suitable for towing heavy loads. The key is understanding the requirements of various jobs and what we like to do – what motivates us – and then developing the skills, the features of the vehicle, and seeking the experiences that match our interests, skills, and motivation. Do that and we will be happy, right?

Well, happiness is not as easy to understand as it may seem. Happiness is not getting the things we want. Psychological research shows that once we get what we want, our happiness sooner or later, and more likely sooner, wanes as we become accustomed to what we acquire. Therefore, it has been said that happiness occurs on the way to fulfillment, and that success is a journey not a destination.

Dr Martin Seligman, former president of the American Psychological Association, believes that there are five factors that contribute to our happiness:

- Positive emotion/pleasure
- Achievement
- Relationships
- Engagement
- Meaning

Therefore, to build the type of career that will make us happy we should not focus solely on our achievements, but that is a useful place to start.

Consider, for example, financial goals. We all know there is more to life than money, but money is important. So how much money do we need to be happy? The answer depends on the purpose of our life. Without a defined purpose, we will not know when we have enough money to achieve that purpose, and we will never be financially 'happy.' How many papers must we write, how many promotions must we get? If we measure these in isolation, we will never have enough, we will never be happy. However, if we are clear about the main purpose in our lives, then we can better judge how much of this or that is sufficient.

Even with clarity in purpose, to achieve extraordinary results the people surrounding us should support our goals. Because no one succeeds alone and no one fails alone, it is important to pay attention to the people around us. We will be happier, and our careers will be better if we spend more time with the right people. But do not confuse this with picking the right people to 'help us.' It is a two-way street, and as we will see in Chapter 7, helping others, coaching, mentoring, etc., can contribute a great deal to our success and to our happiness. It is also important to recognize that the impression others have on us will be influenced by our relationships with them. Although the focus of our careers may not be on making others feel good, it is nonetheless true that when we walk out of our workplace for the last time, people will not be thinking about how many papers we wrote or how many successful projects we worked on. They will be thinking about how we made them feel.

Successful people tend to be clear about their role in the events of their life. Successful people realize no one succeeds on their own. Notable people, or people who receive a lot of attention, may claim to be the 'one' who got it done. However, getting attention is not being successful. In the book *The One Thing* (Keller & Papasan, 2012), the authors quote research showing that the most important difference between less accomplished and elite performers is that the elite performers seek out teachers and coaches and engage in supervised training, whereas the less accomplished tend to try to do it on their own.

Being successful does not mean we were smart enough and good enough to do it on our own. Being successful requires being smart enough and good enough to seek out and work with others who make us and our work better. We will be more successful and happier in the long run when we recognize our success is a shared success with those who helped us, with those with whom we worked.

1.5 REAL-WORLD EXAMPLE

Athletics can be a useful metaphor for life and work, and I have learned a lot about life and work through athletics. However, in the following real-world example, I will not recount my specific experience, but rather use a generic example that is representative of my overall experiences.

Let's apply the success is a journey quote to careers by recasting the quote as: Success in careers is a journey, not a destination. To apply this quote, first think of goals as mileposts on our journey, not the destination. As noted in the previous section, once we achieve a goal, such as a promotion or a level of income, we soon adapt to the new situation. Therefore, achieving a specific goal often does not have lasting, stand-alone satisfaction. Goals are most useful when applied to achieving a greater end.

Second, consider that understanding the route of the journey is more meaningful than the destination. Two people can get to the same place, achieve the same goal, but the journey could be much tougher for one, depending on where they started and what obstacles they encountered along the way. Therefore, the value or reward in a journey is derived not from the destination, but from what is gained along the way. The following example illustrates this point.

Say that you set the goal of running your first marathon. Running 26.2 miles is a difficult task for everyone. So, you train. You and your training partner hold each other accountable, and you meet for your training runs in good weather and bad. You study racing and pacing strategies, practice visualization and motivational techniques, eat a healthy diet, and get proper rest. On race day, you execute your pacing plan, run with great focus and determination, and achieve your goal of finishing the marathon! You earn a coveted finishers medal, and you celebrate your achievement!

But ten years later, what will be important from this experience? You may remember some of the post-race celebration. If you have not lost your finishers medal, you may look at it every few years or so. But you won't do that as a reminder of 'the great race you ran.' You will remember the pride and joy your family shared with you. You will remember the motivation and dedication it took to achieve your goal. You will remember that you had to train smart. You will remember the most difficult training runs with your training partner as clearly as you remember the race.

Awards and rewards are great, but they fade over time. Memories, experiences, friendships, satisfaction, being part of something meaningful, and love, these endure, and these are the mileposts that have proven useful on my journey.

CHAPTER 2

Productivity, Prioritization, and Work–Life Balance

ABSTRACT

Although some statisticians need to get more done to have a successful career, the problem for most of us is that we work too hard. Chapter 2 emphasizes that the point of being productive and prioritizing our work is not to get more done. Rather, the point is to get done what is important as efficiently as possible to have a life outside of work. Work–life balance is critical to having a long, happy, and successful career. Chapter 2 emphasizes that attention management is the key to productivity and that focusing on a few important things rather than trying to get everything done is key to prioritization. Chapter 2 also points out the many distractions in today's work and life environment that make attention management difficult. Chapter 2 includes practical ideas for mastering attention management, controlling

DOI: 10.1201/9781003334286-3

distractions, and prioritizing work, along with real-world experiences from the author that reinforce these points.

2.1 WORK–LIFE BALANCE

Some statisticians would be more successful if they got more work done. However, most of us would have greater potential for long, happy, and accomplished careers if we worked less – and family, hobbies, friendships, tending to our mental and physical health, and other non-work things took on greater importance. A few quotes that help keep things in perspective are:

- The days are long, but the years are short.

- If work is the most important thing in your life, you should work on your life.

- No one says on their death bed they wished they'd spent more time at work.

We can get so caught up in the emergencies and urgencies of the moment, the things that make for 'a long day,' and then our babies are teenagers, our parents have died, we have lost track of cherished friendships, and we are old. Therefore, the points covered later in this chapter on productivity and prioritization are not relevant because they are needed to get more done at work. They are relevant so we can accomplish useful things at work while having time for the more important things in life. As Glennon Doyle points out in her book *Untamed* (Doyle, 2020), it is wrong and misguided to prioritize work over family.

However, independent of family and friends, there is a compelling argument for working less each day in order to accomplish more in the long run. The rested mind is a better mind. A better mind will find better solutions and get more done. Moreover, health problems force unwanted and premature endings to many careers. Despite the time pressures of day-to-day life, making time for our health is a time-efficient approach to careers, not to mention important for our families.

For example, say you work two hours less per week and use that time to get more exercise or to focus on some other aspect of your health. Two hours per week for 50 workweeks per year is an investment of 100 hours per year, which is about 2.5 workweeks per year, or one 'work year' over a 20-year span. Therefore, an investment of two hours per week is time efficient if it helps maintain your health such that you can work one additional year per 20 years in the workforce, or if it increases productivity while working by 5%.

This is a crude calculation. But crunch the numbers any way you can, and the result is always that not prioritizing your health is bad for you, bad for your family, and bad for your career. Therefore, as you consider the next sections on productivity and prioritization, do so with a focus not on getting more work done. Rather, focus on getting the right work done efficiently so that you have time for the more important things in your life.

2.2 PRODUCTIVITY

2.2.1 Introduction

In trying to build a long, happy, and accomplished career, being productive is an essential ingredient. However, in recent years, organizations and individuals have become less productive and have lost critical thinking capacity due to distractions (Newport, 2016). Getting to the root of the problem requires understanding the difference between certain types of work and the different types of focus required to do those tasks.

Deep Work is professional activities performed in a state of distraction-free concentration that push cognitive capabilities to their limit. In contrast, Shallow Work is the non-cognitively demanding, logistical-style tasks, often performed while distracted. Shallow Work tends to not create much new value and is easy to replicate (Newport, 2016). If all we do is Shallow Work, we are easy to replace.

Bailey (2018) defines Hyperfocus and Scatterfocus, which unlike Deep Work and Shallow Work are not contrasting opposites. Bailey defines Hyperfocus in much the same way as he does Deep Work. However, Scatterfocus is a structured mind-wandering that can be an important part of creativity, and therefore is unlike Shallow Work because it does create value. Understanding these types of focus and how to manage distractions is the key to becoming more productive while keeping work hours to levels that allow for work–life balance. The challenges in maintaining work life balance are illustrated in Figure 2.1.

2.2.2 Attention Management

Deep Work is important, and it is difficult. Deep Work is needed to learn and to be productive. However, the amount of time spent in Deep Work is decreasing, even though the outputs from Deep Work are becoming more valuable. Recent trends in business and everyday life decrease our ability to concentrate, to perform Deep Work. Texting, instant messaging, open offices, and meetings make it more difficult to engage in Deep Work. Social media and infotainment may be the biggest distractors because they are designed to hold our attention – that is, to keep us distracted.

Newport (2016) explains the neurocognitive basis for, and consequences of, distractibility. Our brains are wired for distraction. With each distraction (email notification, each check of Instagram or Facebook), we get a spurt of dopamine, the pleasure neurotransmitter. This relationship is depicted in Figure 2.2. Initially, the dopamine increase is in response to the stimulus (e.g., checking Instagram), but over time the dopamine increase is anticipated and occurs in anticipation of the stimulus, not because of the stimulus. In other words, we get rewarded for checking our phone, regardless of what we find when we look.

Simply put, we are rewarded for distraction. Such a minor distraction as the ping from an email notification may seem insignificant. However, distractions leave attention residue. That is, the effect of the distraction remains after the distraction ends. It may take only a few seconds to glance at an Instagram post, but

FIGURE 2.1 Illustration of the challenges in maintaining work–life balance.

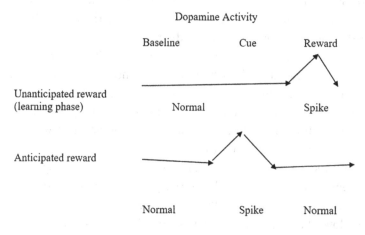

FIGURE 2.2 Dopamine response in the brain to unanticipated and anticipated rewards.

it will take many more seconds for our brains to refocus on the task at hand (Newport, 2016).

Worse still, over time distractions make us more distractible. Doing complex tasks with distractions fires too many neuro-circuits, which in turn reduces myelination of those nerves, thereby reducing neuronal effectiveness. Over time, distractions reduce the ability to focus even when not distracted. In contrast, focused attention fires only the relevant neuro-circuits, which increases myelination and neuronal effectiveness. Over time, concentration increases the ability to focus (Newport, 2016).

Workers have become more distracted. But if distraction is so detrimental to productivity, why do we allow ourselves to be distracted? Why are we working differently today? Newport (2016) posits that as the knowledge economy has gotten more complex, it has become more difficult to measure an individual's productivity and performance. That is, the impact of Shallow Work is hard to detect. Without clear feedback to the contrary, we tend toward behaviors that are easiest at the moment. We fire off quick emails or quick responses to emails (instant messages, or texts). These emails often kick off long email threads instead of well-thought-out emails and responses that resolve the issue. Hence, distractions can propagate more distractions.

Instant communications such as texts and instant messages (IMs) can lead to lack of planning and thought. We fire off a vague question, to which we get back a prompt (aka not-well-considered) response. We have an answer to our question. We got something done. But what did we accomplish? Did we get a 'good' answer? Did we ask the right question?

In this Shallow Work environment, activity is a surrogate for accomplishment. It feels like we are getting more done the longer we spend doing 'stuff.' But we are spending more time doing Shallow Work, multitasking with quick responses to 'urgent' things that are often not that important. We gather and pass along information via meetings, emails, and texts. We stay in contact 24/7 ready to spring into action with more quick responses to things that seem urgent at the moment. In Shallow Work, less time is spent thinking, concentrating, focusing, and creating new information and value.

To understand how big a problem lack of Deep Work is, consider the findings in Table 2.1 (quoted in Newport (2016) for the average day for an average worker).

TABLE 2.1   Daily Activities of a Typical Worker

Checks Facebook 21 times per day

Changes apps 566 times per day

Average duration between interruptions in thought = 40 seconds

Watches five hours of TV per day

1. Data from a business entity wherein the researchers were allowed to monitor the activities of the workers and workers were aware of the monitoring

TABLE 2.2   Four Types of Activities

	Attractive	Not Attractive
Productive	Purposeful	Necessary
Unproductive	Distracting	Unnecessary

That is not the profile of a productive person. Changing apps 566 times means that on average workers were spending 40 seconds on a task. No one can do complex tasks or get a lot of work done by working in 40-second increments. When I first joined the pharmaceutical industry, I thought time management was the most important factor in being productive. Today, it is attention management. Bailey (2018) notes that attention should be managed to achieve our intentions. As a first step in understanding attention management, Bailey (2018) advocates considering the four types of activities as noted in Table 2.2.

Activities are categorized by whether the task is pleasurable or not (attractive vs. not attractive) and whether the activity is productive or not. Productive tasks that are attractive are purposeful activities – where we get the most long-term satisfaction and reward. Time spent on purposeful tasks is rewarding and we are motivated to schedule time for these tasks. Tasks that are unproductive and unattractive are unnecessary. We do not have to manage these tasks because we are not motivated to pursue them.

Productive tasks that are not attractive are termed necessary. We will not be as motivated for these tasks as purposeful tasks, and we will not be as likely to make time to do them. Hence, blocking time to focus on necessary tasks is helpful. Attention management is useful for attractive but unproductive tasks; these are the distracting activities we should work to avoid. Tips for managing attention are covered in the section on putting the principles into practice.

Deep Work as defined by Newport (2016) and Hyperfocus as defined by Bailey (2018) are essentially the same thing, where the focus is on a specific task, such as coding, writing, or solving a math problem. However, Shallow Work as defined by Newport and Scatterfocus as defined by Bailey (2018) are different. Shallow Work is unproductive or less productive tasks. In contrast, Scatterfocus is a structured mind-wandering to connect dots, solve problems, and develop new ideas. That is, Hyperfocus is central to productivity and Scatterfocus is central to creativity and innovation.

In order to sustain periods of intense focus, whether that is Deep Work or Scatterfocus, we need to recharge, to refresh our mental energy. The characteristics of a refreshing work break are:

- Low effort and habitual

- Something you want to do

- Something that is not a chore

Breaks should involve something that is pleasurable and doesn't take much effort, such as a short walk outside, recreational reading, listening to music, a podcast, or an audiobook, spending time with people we enjoy. The more effort put into regulating behavior – to resist distractions and temptations or push yourself to get things done – the more often the need to recharge.

Scatterfocus can be considered connecting the dots. The more dots we collect and the more valuable the dots we collect, the more the information available to be connected. The general categories of dots are:

- Useful

- Balanced

TABLE 2.3 Tactics to Foster Hyperfocus and Scatterfocus Thinking

Disconnect from the internet between 8:00 p.m. and 8:00 a.m.
Walk outside
Do not use your phone as an alarm clock so you are not distracted by it through the night or when you first wake up
Make yourself bored for five minutes and notice what thoughts run through your head
Tame distractions and simplify your environment
Exercise without music or a podcast

- Entertaining
- Trashy

As a rule, consuming more *useful* information is better, especially when we have the energy to process something complex; consume *balanced* information when we have less energy; consume *entertaining* information with intention or when we are running *low* on energy and need to recharge; and consume less trashy information. Of course, it is important to realize that because interests vary, what is trashy to one may be entertaining to others.

To increase the quality of information collected, evaluate what is consumed and try to consume more valuable information. This can be done by focusing on developing the skills and knowledge we find entertaining and by limiting the amount of trash consumed, regardless of whether our definition of trash is the same as someone else's.

The Hyperfocus and Scatterfocus required for Deep Work will not happen on their own. To make Deep Work more habitual, consider the tactics as noted in Table 2.3.

2.3 PRIORITIZATION

2.3.1 Introduction

This section is based on the book *The One Thing* (Keller & Papasan, 2012). The key message from the authors is again a message about focus. Deep Work is about focusing our minds. *The One Thing* is about focusing our minds on what is most important. It may be impossible to emphasize only ONE thing, but narrowing our focus will be useful nonetheless. Keller and Papasan (2012) advocate that a narrower focus to have a bigger impact on fewer things will often lead to the most total impact. This is a reassuring idea to keep in mind for work–life balance. We don't always have to get everything done at work in order to be a useful worker. The authors further consider the major impediments to productivity that result from a failure to prioritize. Those impediments are highlighted in the following sections.

2.3.2 We Must Get Everything Done

Don't you feel good when you get everything done? Most of us do, but this can be an impediment to prioritization, and therefore can be an impediment to productivity and work–life balance. When everything must get done, then everything feels urgent and important. We 'fail' if one thing goes undone. The tendency is to jump into Shallow Work mode, to become active and busy, but often without accomplishing important things. As Keller and Papasan (2012) note, busyness doesn't take care of business.

Many of us use to-do lists to help organize our day. But making a sequence of to-do lists day after day is not a career plan. We should also map the to-do lists to big-picture goals to guard against hopping from one urgent matter to the next without understanding what is most important in the long run. When it gets difficult to move past the urgency of the moment to focus on a long-term goal, remind yourself that the majority of what you achieve will come from the minority of what you do. Extraordinary results are created by fewer actions than most realize. If your to-do list is lengthy, make sure to identify a few critical tasks and the most essential, ONE, task. There will always be just a few things that matter more than the rest, and out of those, likely one will matter most.

2.3.3 Multitasking

No matter how intuitive it is to believe multitasking gets more done, it does not. When we try to do two things at once, we cannot focus on either. Multitasking is often Shallow Work. We can do two things at once, but we cannot do two important things at once, or at least we cannot do them well.

Multitasking can make it feel like we are getting more done, but we are wrong. Keller and Papasan (2012) quote the result of a study that estimated productivity is reduced by 28% in an average workday due to multitasking ineffectiveness. That is, on average, if we did not multitask, we would get as much done in four days of work as we do in five days with the typical amount of multitasking.

When we try to do two or more things at once, we divide our focus and our outcomes are often dumbed down by this shallow thinking. When we switch from one task to another, we activate the 'rules' for whatever we are about to do, while at the same time not having cleared the previous task from our minds. This inherent cost of task switching is not easy to recognize. Therefore, we perceive the benefit of multitasking, but we do not know the cost. We do not know the cost is greater than the gain, so we work very hard for five days to achieve four days of accomplishment.

Newport (2016) explains that the cost of switching tasks is due to attention residue. For up to a minute or two after being momentarily distracted – or after switching tasks – our brains cannot fully engage in the new task, thereby hindering Deep Work. Hence, there is a clear connection between productivity and prioritization. To be productive, we must focus, and that focus must be on the most important tasks. Prioritization is how we create the opportunity for the focus needed to be productive.

2.3.4 Willpower

Keller and Papasan (2012) note that willpower is linked to motivation. When we are motivated, we are willing to do hard things, we have willpower. However, we need to understand how willpower works or we will set ourselves up for failure and disappointment.

No one can summon willpower on demand all the time. Think of willpower (determination, focus) as aspects of mental capacity and draw analogy to physical capacity. As we do intense physical work, we fatigue and our capacity decreases. No amount of willpower can allow a runner to sprint at the start of a marathon and still finish strong. No amount of willpower can allow a runner to run 26 miles day after day. Strategies based on this hope are bound to fail. Marathoners know that proper pacing, proper rationing of their energy, in training and racing, along with recovery between hard efforts, is key to success. Similar ideas apply to mental capacity, willpower, and focus.

Intense mental work causes fatigue and decreases mental capacity. No amount of willpower can overcome that fatigue. Remember, the three key navigation points introduced in Chapter 1: Are we motivated; are we working smart; and are we working well with others. We need to work smart, or we will waste our motivation.

Recognizing mental fatigue helps us to prioritize our day. Keller and Papasan advocate that we should do what matters most when our willpower and mental capacity are the highest. To get the most out of the workday, do the most important work when willpower is not drawn down. Keller and Papasan (2012) highlight the idea of dividing work into two general buckets: Maker (do or create) and manager (oversee or direct). 'Maker' time requires large blocks of the clock to write papers or code, develop ideas, generate leads, recruit people, produce products, or execute on projects and plans. This time tends to be viewed in half-day increments. 'Manager time,' on the other hand, gets divided into hours. This time typically has one moving from meeting to meeting, or task to task, and because managers tend to have power and authority, the meetings happen at their cadence.

The idea applies even to those who do not manage others. Think of manager time as managing your own affairs. Manager time can be used to catch up on emails, meetings, or any task that is less cognitively demanding. For the morning people, do the most important tasks early in the day. For night-owls, block time for the most important tasks late in the day. But remember, no matter how hard we try, there will always be things left undone at the end of the day, week, month, year, and life. Trying to get everything done is counterproductive.

2.3.5 Identifying What Is Most Important

To identify what is most important, the top priority, Keller and Papasan (2012) advocate the following: Ask what is the one thing that, if done well, will make everything else easier and better. Consider adding a time frame – such as 'right now' or 'this year' to give your answer the appropriate level of immediacy; or, 'in five years' or 'someday' to find a big-picture answer or long-term goal. Say the category first, then state the question, add a time frame, and end by adding 'such that by doing it everything else will be easier or unnecessary.' For example: 'For my job, what is the ONE thing I can do to ensure I hit my goals this week such that by doing it everything else will be easier or unnecessary?'

We may not always be able to limit our answer to one thing. However, even narrowing the list to a few things can focus our work. To help make the idea of focusing on one thing more tangible, here are some examples:

> The one thing I must do this week to ensure I hit my goals is to complete the simulations needed for sample size determination in study WXYZ.

The rationale might be that other functions needed to know the sample size as soon as possible to initiate other study start-up tasks. If we get the sample size info needed by others, we can then have more distraction-free time for other tasks.

The one thing I must do this year to ensure I hit my goals is to complete my self-study of making decisions under uncertainty because I will not be ready for broader roles and leadership opportunities until I improve my skill in making critical decisions.

2.4 PUTTING THE PRINCIPLES INTO PRACTICE

Attention management is the key starting point for increasing productivity. Newport (2016) proposes four options to increase Deep Work that involve establishing routines and work habits to help focus attention. The best approach depends on a variety of factors. The four options are summarized in Table 2.4.

Many may find the rhythmic Deep Work strategy most applicable. Some key tactics to foster rhythmic Deep Work are listed in Table 2.5.

Despite our best efforts, distractions are inevitable. Therefore, the better we can manage distractions, the more Deep Work we can do. We can further manage attention by scheduling Deep Work for parts of our day when we are mentally freshest. To manage distractions, consider the categorizations in Table 2.6. Distractions are categorized as fun vs not fun and can vs cannot be controlled.

TABLE 2.4   Four Approaches to Deep Work

Approach	Implementation
Monastic	Radically minimize Shallow Work in all aspects of life
Bimodal	Schedule long stretches of time for isolation and Deep Work (days or weeks)
Rhythmic	Schedule time every day to do Deep Work – i.e., manage attention/distractions
Journalistic (very hard to switch like this)	Switch into Deep Work whenever the opportunity presents

TABLE 2.5   Tactics to Foster Rhythmic Deep Work

Limiting social media and infotainment
Limiting availability on IM/texts
Turning off email notification and batch process emails
Limiting multitasking
Use quiet places away from your workstation for focus time

TABLE 2.6   Tactics for Managing Distractions

	Fun	Not fun
Can control	1	2
Cannot control	3	4

For distraction category 1, fun distractions that can be controlled: Control these distractions by setting limits in advance. For example, check Facebook for x minutes during lunch. For distraction category 2, activities that can be controlled and are not fun: Control these by planning to avoid the distraction. For example, when working on a manuscript plan to work from home to avoid distracting conversations with co-workers in an open office setting. For distraction category 3, fun activities that cannot be controlled: Enjoy the moment, it is fun. But be ready to refocus. When you unexpectedly get a call from an old friend, take the call, but do not go on forever, and when you hang up, refocus on your task. For category 4, distractions that cannot be controlled and are not fun: Develop a coping mechanism. For example, wear headphones to block noise and to signal to others you are focusing.

Because Deep Work is hard, schedule it for when you have the most energy. During extended blocks of Deep Work, take short breaks. Research suggests it is optimal to take 17 minutes off after 51 minutes of work, or others say 15 minutes off every 90 minutes. I have had more success with a 50–10 rule, or a 55–5 rule, 50 (55) minutes work, 10 (5)-minute break, because it fits into typical office schedules and is more practical, even if not optimal.

It is tempting to ignore the 50–10 rule because the best way to get the most done in the next hour is to work 60 minutes, not 50. When traveling, the best way to get the furthest is to drive for 60 minutes. But if you bypass a gas station in that hour and run out of fuel, working harder in the previous hour (driving further) means less progress in the long run. Getting the most done each hour of each day is not the best way to get the most done in the long run. Running all out at the start of a marathon is not going to win the race.

It is possible to conflate Deep Work with working alone. Deep Work should not come at the expense of collaboration. Consider the following approach for Collaboration and Deep Work. Utilize a real or hypothetical hub and spoke layout as depicted in Figure 2.3. In hubs, interact with people and gain exposure to ideas and approaches. Then use Deep Work approaches in locations where distractions are minimized (spoke) to work on what was encountered in the hub.

Examples of hubs could be any gathering place for conversation or even formal meetings such as meeting rooms or designated collaboration spaces. Examples of spoke locations could be your desk if it is in a non-distracting environment, or work from home, or work in a quieter location away from your desk at the office. In considering remote/hybrid work models, the office could be the hub and the home office could be the spoke.

To manage our day(s), week(s), and year(s) to generate more Deep Work, we can prioritize. Prioritization provides the opportunity for focus. A common impediment to prioritization is difficulty in saying no. Saying no is hard; we do not want to disappoint our boss or co-workers. However, when we say yes to something, we cannot manufacture the time to do it. There is something else that must be left undone. Hence, it is imperative to say no sometimes so that we can focus on the most important things. Saying yes to everyone can be similar to saying yes to nobody as we lurch from one new priority to another. Remember, too many changes in direction imply no direction at all. When considering whether

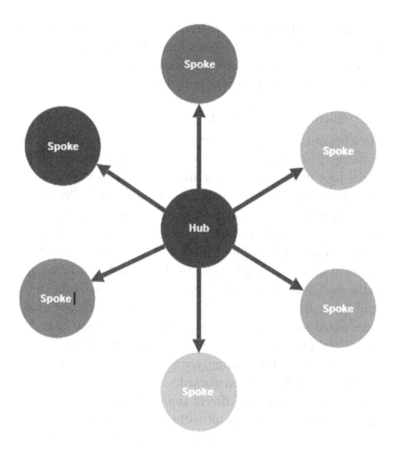

FIGURE 2.3   Example of hub and spoke layout to facilitate Deep Work and Collaboration.

to say no, remember the main idea is to narrow focus so that we can have a bigger impact on fewer things. We are trying to build success one big thing at a time, not by doing everything at once.

2.5 REAL-WORLD EXAMPLES

The following are useful quotes that reinforce the ideas presented in this chapter

> I can do 12 months of work in 11 months, but I cannot do 12 months of work in 12 months.

This quote reinforces the need to recharge. We are often scared to unplug; we fear we will miss out on something important. But remember, most of our contributions come from a few important things rather than hundreds of little things.

2.5.1 Attention Management

I first became aware of the merits of short breaks while trying to write complex computer programs as a graduate student. Many times, I got stuck and could not find the bug or logic error in my code. My frustration level would rise, I would work harder but no matter how hard I stared at the screen I was often unable to find the problem. Frustration would increase and I would push back from my chair and head for the student center to get a snack or a soda. It was less than a 5-minute walk. Often, by the time I got to the student center, I would have an idea of how to fix the code. And it usually worked. I had to learn the hard way that brute force would not fix the problem. I needed to get away from the screen, get some fresh air, and reset my thinking so that I could step back, process information, connect the dots, and produce a potential solution that often worked.

Over time, I adopted different approaches to attention management, but most of them fit within the rhythmic category – finding stretches of time for Deep Work within the day. Before working from home was a common option, I shared a ride to work with a neighbor. He left VERY early. But this allowed me to get in a couple hours of focus time to write or code before others showed up at work and before meetings began. When writing this book, I was working for a Biotech company based in San Francisco, while living in Indianapolis. Since I was in the Eastern time zone and most of my co-workers were in the Pacific time zone, I could count on a few hours of undistracted writing time in the morning.

In general, the larger the organization I worked for, the more explicit I had to be in blocking focus time. In my experience, larger organizations tended to have more meetings and more non-core responsibilities. In those situations, if I did not block time on my calendar the time always seemed to evaporate.

In the academic settings in which I worked, it was possible to utilize the bimodal approach, although I was not aware of Deep Work principles at the time. For example, as a graduate student, it was possible to block entire days or weeks to focus on qualifying or final examinations, or to work on my thesis or dissertation. In my initial academic appointment, the time between semesters when I was not teaching was a good opportunity to employ the bimodal approach to focus on large research projects.

2.5.2 Prioritization

Some of the best advice I ever received on prioritization is summarized by the following quote:

> Keep your eye on the football.

The 'football' is the shape made by the overlap of circles in a Venn diagram. The point is to figure out what is important to your team or organization and figure out what you like to do or what is important to you. Then spend as much time as possible where these two spheres overlap.

It is not always so easy to understand the priorities of your team or organization. Therefore, it is worthwhile to discuss priorities with your supervisor to avoid misunderstandings. In so doing, do not just discuss what is at the top of the priority list, also clarify what is least important. The following is a contrived example, but it illustrates some of the most useful prioritization discussions I have had with my supervisors:

> Thanks for clarifying our priorities, Janelle. It is clear now that project X is most important. And I can also see how projects Y and Z fit into the overall picture. Given my understanding of the timelines, we may encounter some bottlenecks. If necessary, would it be appropriate

to let timelines slip a bit on Project Z in order to ensure delivery of Project X?

The key point here is that it is often easy to say what is most important, but it is less clear what may need to go undone in order to achieve the most important goals. If you are a supervisor, a similar conversation may go as follows.

Jim, our top priority is Project X (INSERT BRIEF EXPLANATION IF NEEDED). Projects Y and Z are also meaningful, but not as critical as X. So, Jim, please manage your time to focus first on Project X. Given my understanding of your workload, I think it will also be possible to support the timely delivery of Projects Y and Z. However, this is something we will have to keep a close eye on. Please let me know how things are progressing. If things start to slip on Project Z, we will need to consider options, such as pulling in additional support or changing the timeline. And, as you manage your time, remember, Project X is very important and we need your best work, not your most work.

CHAPTER 3

Creativity and Innovation

ABSTRACT

This chapter uses a broad definition of innovation and creativity, not just inventing something new, to introduce key concepts in what leads to innovation and creativity and how to solve tough problems. Through understanding key principles in what fosters innovation and creativity, we see that although some people are inherently more creative and innovative than others, each of us can increase our capacity for innovation and creativity. As with productivity, there is an element of attention management that is key to fostering our innovative and creative capacity. This chapter includes practical advice on how we can manage our attention and expose ourselves to new ideas and creative environments to increase our capacity for innovation, creativity, and solving tough problems. These practical ideas are reinforced by real-world examples from the author's career.

DOI: 10.1201/9781003334286-4

TABLE 3.1   Examples of Golden Nuggets Statisticians Can Find

Creative/innovative idea or approach
Applying known theory in a novel way or context
Solution to a tough problem
Making or contributing to a complex decision
Leading a large and/or complex task
New analytic approach/data mining plan
Better design/development plan
Better programming approach
Finding an influential result
Finding a better way to explain something complex
Key coaching/mentoring advice

3.1 INTRODUCTION

Where do good ideas and innovation come from? How can we produce more good ideas? Creativity and innovation are not static traits fixed at birth. Yes, some people are more creative, more innovative, and have more good ideas than others. However, regardless of inherent ability, much can be done to enhance creativity, to produce more good ideas, and to find more solutions to tough problems.

Innovation and creativity can mean different things to different people and in different contexts. For our purposes, think of innovation and creativity as being broader than inventing a new device or a new analytic method. I like to use the term finding golden nuggets as a definition for innovation and creativity. Table 3.1 lists some of the golden nuggets statisticians may find. The ideas and material covered in this chapter can help us find more golden nuggets.

3.2 UNDERSTANDING INNOVATION AND CREATIVITY

3.2.1 Overview

Where do good ideas come from? Steven Johnson's (2011) answer in *Where Good Ideas Come From* is coffee houses, or wherever people and their ideas gather or collide. His more detailed answer is that most ideas come from what he terms the adjacent possible and from six general tactics for assembling the building blocks of ideas and innovations.

Before diving into those ideas, consider the following from Grant (2016): Idea quantity encourages idea quality. Although it is often said that quality is more important than quantity, regarding ideas, quantity and quality are both important. The quantity of ideas encourages the quality of ideas during brainstorming through divergence in thinking. Grant (2016) quotes research from psychologist Dean Simonton that supports this point. Simonton found that creative individuals do not produce a higher percentage of successful ideas, they produce more ideas. Increased idea output leads to a higher probability of developing a good idea. The key to producing a good idea is to have more ideas.

Grant (2016) illustrates this point using the career of Pablo Picasso. Picasso's work included 2,800 ceramics, 1,800 paintings, 1,200 sculptures, and more than 12,000 drawings. But he is famous for a few quality ideas. Simonton provided further support in reporting that geniuses could not tell which of their works would become a timeless classic and which would flop. Therefore, it is best to produce many ideas to increase the probability of success.

Although Grant (2016) argues that the quantity of ideas is essential, he does not advocate trying to churn out ideas on an assembly line. Instead, he advises that we should give ideas time, which is further reinforced by the slow hunches idea of Johnson (2011) discussed in *Where Good Ideas Come From* (covered in the next section). Hence, do not immediately reject any idea, but do not throw out random thoughts as a replacement for well-conceived ideas.

Grant's (2016) and Johnson's (2011) ideas of slow hunches, giving ideas time, are based on the recognition that most good ideas do not stem from a 'eureka moment.' Most good ideas are the product of gradual progress rather than an instantaneous break-through or even rapid progress. This idea is covered in more detail in a subsequent section. However, the key point is that going slow provides time for broad and deep thinking, to consider a wider range of ideas before settling on a decision. Sometimes it is good to set an idea aside for a time and forget about it. Leonardo da Vinci started painting the Mona Lisa in 1503, but then abandoned the project to work on something else. He finished the Mona Lisa in 1519, 16 years later. During this time, da Vinci experimented with optical illusions and new painting techniques. By using this experimentation time, he returned to the Mona Lisa with a different mindset and new ideas.

Johnson (2011) uses analogies from evolution to explain how and why innovation thrives in collaborative networks where opportunities for serendipitous connections exist. The following sections summarize those ideas.

3.2.2 The Adjacent Possible

Johnson (2011) notes that evolution and innovation seldom happen in huge leaps. Small increments of innovation are more common – because of the adjacent possible – an idea he explains using the following example from evolution.

Four billion years ago, carbon atoms wandered in the primordial soup. As life began, those atoms did not spontaneously arrange themselves into complex life forms. First, they had to form simpler structures like molecules, polymers, proteins, cells, and primitive organisms. Each step opened possibilities for new combinations, expanding what was possible until a carbon atom could reside in a sunflower or other complex organism.

Johnson (2011) notes that eBay could not have been created in the 1950s, no matter how brilliant the scientists and entrepreneurs of the day. First, someone had to invent computers, then a way to connect those computers, then a browser, and then a platform for online payments. Both evolution and innovation tend to happen within the bounds of the *adjacent possible*, within the

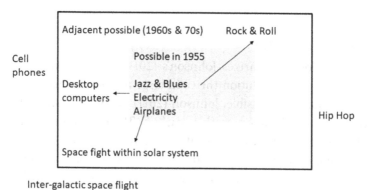

FIGURE 3.1   Illustration of the adjacent possible.

realm of possibilities available at that time. See Figure 3.1 for an illustration of the adjacent possible.

Johnson (2011) points out that great leaps beyond the adjacent possible are rare and are typically short-term failures. The environment is not ready for them. Consider the difference engine, envisioned as an automatic mechanical calculator to tabulate polynomial functions, designed in the 1820s by Charles Babbage. The name, the difference engine, is derived from the method of divided differences, which is a way to interpolate or tabulate functions by using a small set of polynomial coefficients. Some of the most common mathematical functions used in engineering, science, and navigation were, and still are, computable with the use of the difference engine's capability of computing logarithmic and trigonometric functions that can be approximated by polynomials. Hence, a difference engine could in theory compute many useful tables of numbers.

Thus, in the 1820s, Babbage had outlined the first computer. However, the metallurgy of the day could not support the precision of the gears, and the cost and size of the difference engine proved prohibitive. Vacuum tubes made high-speed calculations possible, and semi-conductors made them routine. The difference engine was too big a leap for the day, even though the idea was good.

3.2.3 Liquid Networks

This section summarizes Johnson's (2011) explanation of why innovation and evolution thrive in large networks where many connections are possible. Johnson (2011) begins by noting that the basis of all life on earth is carbon. The electron structure of carbon is uniquely good at connecting with other atoms (covalent bonding) and can thus construct complex chains of molecules. Carbon is especially good at connecting with hydrogen, oxygen, nitrogen, and phosphorus. Over 99% of all matter on earth is made of these atoms. These connections allow new structures like proteins to emerge – the adjacent possible. Without carbon, the earth would have likely remained a dead soup of chemicals. Combinations of carbon made the phospholipid structure of cell walls, which in turn allowed certain things to move in and out of cells. With cells, new adjacencies emerged and life evolved.

Johnson (2011) extends the carbon idea to connections that facilitate ideas. When humans first began to organize into settlements, towns, and then cities, they became members of networks. The network exposed them to new ideas and allowed them to share their own. Before towns, a novel idea by one person could die with her because she had no network to spread it. That is why great ideas are more likely to arise and spread where people gather.

Johnson (2011) cites a study geared to better understand the roots of scientific breakthroughs. In the 1990s, researchers recorded everything that went on in four molecular biology laboratories. It is reasonable to expect important discoveries in molecular biology to arise from individuals peering through a microscope. However, the most important ideas arose from regular lab meetings where the scientists discussed their work in an informal setting.

Other studies have shown that more creative individuals tend to have broad social networks that extend outside their own organization, and hence can get new ideas from different contexts.

Cities facilitate large networks that allow ideas to be diffused and combined in novel ways. This is one of the reasons why cities are more creative than smaller towns. Today, the World Wide Web is an enormous network, creating, connecting, and diffusing more ideas than any network before it.

3.2.4 Slow Hunches

Johnson (2011) explains that big ideas tend to evolve over time as slow hunches rather than sudden breakthroughs. He states that the eureka moment is not a myth, but it is a misunderstanding of how the brain is working. In retrospect, great discoveries may seem like single, definable eureka moments, but that 'flash' is just the moment when the final connection was made between thoughts that were slowly evolving and coalescing over time. Ideas are often gradually maturing *slow hunches*, which take time and effort to develop. A hunch is not some vague entity in the ether, it is a connection of neurons that form an idea in the brain. Hunches and ideas are like a garden; they benefit from the opportunity to grow and mature, and will grow and mature best when tended.

Darwin said the theory of natural selection popped into his head when he was contemplating Malthus' writings on population growth. However, Darwin's notebooks document that long before the so-called epiphany, he had described the key tenets of natural selection. This slow hunch matured into a fully formed theory over time. In retrospect, an idea may seem so obvious that it must have come in a flash of insight, but the key ingredients were building over time.

As a child, Tim Berners-Lee read a Victorian-era how-to book and was fascinated by the 'portal of information' he had found. Over a decade later, working as a consultant at the Swiss CERN laboratory and in part inspired by the book, he tinkered with a side project, which allowed him to store and connect chunks of information, like nodes in a network. Another decade later, CERN authorized him to work on the project, which matured into a network where documents on different computers could be connected through hypertext links. After decades of developing and maturing, the World Wide Web was born in a flash.

3.2.5 Serendipity

Johnson (2011) explains how connections between ideas drive innovation. Those connections often seem to be from chance; however, increasing the opportunity for connection increases the probability of connections that foster innovation. As an evolutionary example, Johnson (2011) notes that carbon being able to bond, to form connections, with other atoms was vital for evolution, but a second, randomizing force was also necessary: Water. Water moves and churns, dissolving and eroding things in its path, thus fostering new connections between atoms in the primordial soup. Just as important, the strong hydrogen bonds of water molecules helped maintain those new connections.

This mix of turbulence and stability is why *liquid networks* are optimal both for the evolution of life and for creativity. Innovative networks teeter on the brink of chaos, like molecules in churning water. Random connections drive serendipitous discoveries. Dreams can be considered the primordial soup of innovation, where ideas combine without clear connections. In fact, neuroscientists have confirmed that 'sleeping on a problem' does help to solve it. Centuries ago, the German chemist Kekulé dreamt of a mythological serpent devouring its own tail. Later, he realized how carbon atoms in a ring, like a serpent eating its tail, formed the molecule benzene, thus solving one of the great mysteries of chemistry that had befuddled scientists for decades.

Johnson (2011) explains that chaos and creativity are linked on a neurological level. Ideas are manifestations of a complex network of neurons firing in the brain, and new ideas become possible when new connections are formed. For some reason, neurons in the brain alternate between states of chaos, where they fire out of sync with each other, and more organized *phase-lock* states where large clusters of neurons fire at the same frequency. The duration spent in either state differs from brain to brain. Studies have shown that longer spells of chaos tend to be seen in smarter persons.

Serendipity should not be confused with pure luck. There are things we can do, such as in some way being part of a network, to enhance the probability of a lucky connection.

3.2.6 Errors Can Lead to Innovation

Johnson (2011) examines how great innovations have emerged from environments that are contaminated by error. He cites natural reproduction, where genes are passed from parent to offspring, providing 'building instructions' for how the offspring should develop. Without occasional mutations, evolution would have come to a virtual standstill. The elephant's tusks or peacock's feathers would have never emerged if only perfect copies of existing genes had propagated. Mutations create variation, variation creates opportunity, opportunity fosters change. Most mutations fail to take hold in the genome because most mutations convey no advantage. However, some mutations yield advantages and these errors produce winners that drive evolution.

Alexander Fleming discovered penicillin because he allowed a bacteria sample to be contaminated by mold and wondered what had killed the bacteria. In fact, major new scientific theories have begun as annoying errors in the data, which keep demonstrating that something in the dominant theory is wrong.

Unexplained errors force us to adopt new strategies and to abandon old assumptions. Johnson (2011) describes a study in which researchers showed two groups of people slides with various colors on them and asked the subjects to free-associate words after seeing each slide. Here is the twist: Actors (aka research confederates) were inserted into the second group who sometimes claimed to see different colors than the actual one shown, e.g., 'green' when the slide was blue.

The first group came up with only the most predictable associations, e.g., 'sky' for a blue slide. The second group was more creative. The 'error' introduced into the group forced them to consider more possibilities than just the obvious ones.

3.2.7 Exaptation

Johnson (2011) explains that many inventions are from borrowed, repurposed, or connected ideas drawn from multiple disciplines. Evolutionary biologists use the term *expatiation* to describe the phenomenon where a trait developed for one purpose is used later in a different way. Feathers first evolved for temperature regulation, but they evolved into their airfoil-shape that today helps birds fly.

Ideas are often repurposed. Tim Berners-Lee created the World Wide Web as a tool for scholars, but in the course of time it became a network for shopping, social networking, and many other things. Johannes Gutenberg found an innovative use for a 1000-year-old invention: The wine screw press used to squeeze juice out of grapes. Using this ancient technology and his knowledge of metallurgy, Gutenberg created the world's first printing press. Today, Nairobian cobblers make rubber sandals out of discarded car tires. The old is reshaped into the new.

Discarded spaces are also transformed through innovation. Just like the skeletal structure left behind by dead coral forms, the basis of the rich and thriving ecosystem of the reef, abandoned buildings, and rundown neighborhoods are often the first homes of innovative urban subcultures.

The unconventional thinking and experimentation of new activities do not fit in glitzy, high-priced, mainstream malls and shopping districts. Old buildings allow subcultures to interact and generate ideas that then diffuse into the mainstream. And, of course, subcultures can be supported in larger cities better than in smaller towns.

Benjamin Franklin and Charles Darwin favored working on multiple projects in a slow, *multiprocessing* mode. One project would take center stage for days at a time, but linger in the back of their mind afterwards as attention shifted to another project. This enabled connections between projects to be drawn. It is important to note that multiprocessing is different from multitasking. In multiprocessing, focus is on one thing at a time, but ideas are allowed to mingle and to be cross-referenced over time as connections between the partially completed projects mix. However, two projects are not being worked on at the same moment.

The philosopher John Locke understood the importance of cross-referencing as early as 1652, when he began developing a system for indexing the content of his *commonplace book*, a scrapbook of interesting thoughts, findings, and ideas. These books formed his repository of ideas and hunches, maturing and waiting to be connected to new ideas.

3.2.8 Platforms

Johnson (2011) explains how platforms are springboards for innovations. In Ecology, a *keystone species* is one that is important to the welfare of the ecosystem. On a small island with no other predators, a pack of wolves keeps the population of sheep under control, thus preventing the sheep from overpopulating and eating the island bare of forage, which would collapse the entire ecosystem.

Ecologists have further defined a specific and important type of keystone species, the *ecosystem engineers* that create habitats for other organisms, building *platforms* from which others benefit. Consider for example the beavers that dam rivers turning forests into wetlands, or the coral that builds thriving reefs into the middle of the ocean (Figure 3.2).

Platforms are important to innovation too. Johnson (2011) cites the Global Positioning System (GPS) as an example of a platform technology. Originally developed for military use, GPS has led to countless innovations: From GPS trackers to location-based

FIGURE 3.2   Example of a beaver creating wildlife habitat – a platform.

services and advertising. Platforms often stack on top of each other, meaning that one platform provides the foundation for more platforms, which again produce new innovations.

An example is the beaver, who fell trees. The fallen trees attract woodpeckers who drill nesting holes in the dead trees. Once the woodpeckers have left, these holes are occupied by songbirds. Beavers create platforms for woodpeckers, who in turn create platforms for songbirds. The story of Twitter is similar: The Web was based on existing protocols, Twitter was built on the Web, and now countless apps have been designed on the Twitter platform as the adjacent possible is expanded at every step.

Johnson (2011) summarizes his insights for how individuals can increase their chances of generating good ideas and innovation. He states that the patterns are simple, but followed together, they make for a whole that is wiser than the sum of its parts. Go for a walk; cultivate hunches; write things down and don't worry about how it is organized; embrace serendipity; make generative mistakes; take on multiple hobbies; frequent coffeehouses and other liquid networks; follow the links; let others build on your ideas; borrow, recycle, and re-invent.

3.3 CONVINCING OTHERS

We saw in Chapter 1 the sad example of Dr Semmelweis, who was unable to convince the medical community of the benefits of handwashing in preventing infection. Being right is not enough. We must convince others. Chapter 6 covers leadership and influence in detail. This section summarizes ideas from Adam Grant's book (2016) *The Originals* on how to convince others your idea is worth pursuing. Table 3.2 lists those ideas.

3.3.1 Admit Weaknesses

Grant (2016) suggests offering your opinions and ideas in a balanced manner. It can be effective to begin by noting the shortcomings or costs of your idea. People will be more willing to accept the benefits of your proposal if they understand the weaknesses; admitting weaknesses increases our credibility and the legitimacy of our ideas.

3.3.2 Make Radical Ideas Seem More Normal

Grant (2016) notes that humans tend to reject things that are not familiar. This tendency hinders thinking of and accepting

TABLE 3.2   Methods to Convince Others on the Merits of a New Idea

Admit the weaknesses of your idea so others can focus on the merits

Normalize new ideas by gradually exposing the new idea rather than forcing it on others

Collaborate with others who have opposing viewpoints

Shape ideas to gain support. Rather than saying how revolutionary your approach is, show how it builds upon what is already accepted original ideas. Hence, a way to get an idea accepted is to increase familiarity with it. Repeating original ideas can help give others time to warm up to the idea. For example, introduce an original idea, wait a few weeks, and then reintroduce the idea after your coworkers have had time to normalize the idea. Research suggests that consistent exposure to a new idea allows people to become more receptive to that idea over time.

Another tactic Grant (2016) advocates is to integrate an original idea into existing, accepted ideas. This helps others understand the application of the idea, and putting the idea in a familiar context can help make the idea seem more familiar. Grant (2016) provides the example of the Disney film, *The Lion King*. When the original idea was pitched, Disney was skeptical. The producers did not like the dark storyline. However, Disney CEO Michael Eisner changed people's opinions of the film by framing *The Lion King* within a familiar context. Eisner highlighted the similarities between the *Lion King*'s themes and Shakespeare's *King Lear* and *Hamlet*. The producers were persuaded when they were provided a common point of reference that allowed an original idea to fit in a mainstream context.

3.3.3 Collaborate with Opposing Opinions

Group think is discussed in detail in Chapter 5 on critical thinking and making decisions under uncertainty. Group think is a negative phenomenon that occurs when people within a group prioritize avoiding conflict by reaching a consensus on the best choice. Group think was first described as a phenomenon by Yale researcher Lester Irving Janis. Janis outlined group think as one of the biggest drivers of poor team decision-making.

Grant (2016) noted an experiment that further supports the importance of including diverse views on key decisions. A psychologist ran an experiment where groups of participants had to hire one of three candidates. The first candidate was presented as having the best skills for the job. Therefore, (s)he was the clear best choice. In one condition, the psychologist had research confederates show a preference for the less qualified candidate. In this condition, the genuine research participants conformed to the majority position and chose to hire the less qualified candidate. The second condition was similar to the first except just one research confederate offered a minority position by backing the best-qualified candidate. In this circumstance, the probability that the research participants choose to hire the best-qualified candidate quadrupled. The inclusion of just one minority opinion prevented group think.

3.3.4 Shape Your Ideas to Gain Support

Grant (2016) notes the importance of striking the right tone when conveying ideas. He notes the need to keep people interested but cautions against pushing too hard. Horizontal hostility is a form of prejudice that can occur between members of a group. You can avoid horizontal hostility by connecting your idea to existing values.

For example, if you believe your department needs to make a major change in direction, stating your opinion may alienate some. Even if you are right, you may not have enough support to institute change. However, a slower approach, first connecting over the shared ideas, the goals of the team, and then coaxing the audience in your direction, is more likely to be effective. Persistence is important because the initial response to a bold idea may be less than enthusiastic, but responses tend to improve over time.

3.4 PUTTING THE PRINCIPLES INTO PRACTICE

Creativity is not a static attribute. The capacity for creativity and innovation, for finding the golden nuggets, can be increased by understanding where good ideas come from and how the brain works in processing those ideas.

Good ideas and innovations are more likely to come from networks. Therefore, innovative and creative capacity can be increased by exposure to networks, such as groups, blogs, and seminars. Anywhere we can be exposed to new ideas; it can be a network, a springboard to creativity and innovation. Serendipity in being exposed to a new and useful idea should not be confused with pure luck. There are things we can do to enhance the probability of serendipity – to enhance the chances of forming useful connections. We can utilize our networks.

We can take advantage of regular teams and functional interactions and reach further to be part of industry work groups, journal clubs, etc., as appropriate for our interests, goals, and responsibilities. The greatest network is, of course, the World Wide Web, where a wealth of ideas is available and hyperlinked for easy connections between disciplines.

Research has shown that employees who worked across areas solve more problems than those who focus on a narrow area. Therefore, problem-solving capability can be enhanced by working in diverse areas.

Reading is a great way to form new connections. Deep dive reading is useful because we read a volume of material in short order and thus have the ideas in mind more than from readings long ago. As such, it is useful to read on a variety of subjects. If, like most of us, you cannot do deep dive reading, store key ideas from each reading in a 'common book' and revisit the common book. As a specific example, consider the 'Great Courses' series of audiobooks. Each chapter is a lecture topic, 30 or 45 minutes long. Most books have 25–40 chapters. Thirty minutes of listening, five days per week yields 6–10 books per year. Do that for five years – and reading 30–50 books WILL make a difference – and it will make a bigger difference if you store key ideas and revisit them.

Recalling the hub and spoke idea from Deep Work in Chapter 2, networks can be used as the hub for exposure to new ideas and connections. The spoke is where we go to do individual work, including work that follows up on the ideas encountered in the hub. Hence, productivity, creativity, and problem-solving capability can be enhanced by controlling HOW we work.

The idea of Scatterfocus first discussed in Chapter 2 is relevant to creativity and innovation. Scatterfocus is a structured mind wandering, which can be most useful after recent exposure to new ideas. Walking, especially in nature, showering, work breaks, vacations, and working from an unfamiliar location, all provide breaks from the normal routines of life where the phase-lock part of the brain dominates. New and unfamiliar settings result in less time in phase lock, allowing for the formation of more new connections and the cultivation of ideas in the subconscious mind.

Knowing that many ideas arise from slow hunches suggests several approaches to enhance creativity: (1) Interacting with others doing similar things; and (2) making space/time for slow hunches, free from the urgent pressures of today. Good ideas are more likely when we give ourselves time to let the ideas develop, including through refinements generated via interactions with others.

An effective way to tend ideas over time is to write them down. Many great scientists 'indexed' their ideas in a common book. The common book could be a collection of key quotes and ideas from other readings, and/or incorporate one's own ideas from hunches, experiments, and conversations. By writing things down, hunches have a chance to bloom into fully formed ideas.

Slow multiprocessing – NOT MULTITASKING – is a sequential processing of concurrent projects. It is yet another way to form connections as the mind moves across multiple problems to make connections.

3.5 REAL-WORLD EXAMPLES

The following are actual examples from my career where I was successful and unsuccessful in developing and getting new ideas accepted

3.5.1 An Example of Exaptation

Missing data that arise from early discontinuation of patients is a well-known problem in clinical trials today, but it was only an emerging issue in 1998 when I joined the pharmaceutical industry. In the genetic evaluation of livestock, which was a central focus of my graduate training in Animal Breeding and Genetics, farmers and ranchers tended to send in pedigree and performance evaluation data on only those animals they deemed worthy of entering their own breeding herd or that could be sold as breeding stock. Data on the inferior animals that were culled were not always sent to the breed registry for inclusion in the database to avoid having to pay the processing fees. Therefore, in both clinical trials and genetic evaluations, the tendency is that data from those with bad outcomes are more likely to be missing.

In genetic evaluations, we had used likelihood-based mixed-effects models to account for the bias from selective reporting. When I joined the pharmaceutical industry, I was surprised to see that an ad hoc, single imputation approach known as LOCF (last observation carried forward) was popular. Although likelihood-based mixed-effects models were sometimes used as an ancillary analytic approach, the full merit of these analyses was not exploited.

A series of research projects that I was involved in and research from others showed the merits of the likelihood-based repeated-measures approach, a specific version of which we named MMRM. At the company where I worked at the time, MMRM replaced LOCF in many settings and played an important role in the development of several drugs. This is an example of exaptation. I did not invent likelihood-based inference or mixed-effects models. However, borrowing ideas that were well entrenched in genetic evaluations and applying them to longitudinal clinical trials led to an important advancement.

3.5.2 An Example of Working with Others of Differing Viewpoints

Collaboration was essential in establishing the merits of the MMRM analytic approach. When the MMRM versus LOCF research was being conducted, there was a tendency for competition between the likelihood-based group, of which I was a member, and the multiple imputation camp. A common question for those wishing to move beyond LOCF to more principled analytic approaches was: 'Should I use MMRM or Multiple Imputation (MI)?' The correct answer was yes, not one or the other! Through collaboration, the two schools of thought recognized that the conceptual underpinnings of both approaches rested on the same statistical principles. By couching MMRM and MI as complementary methods rather than competing methods, the arguments and motivation for moving away from LOCF were stronger.

3.5.3 Two Examples of Failure to Gain Full Support for a New Idea

In the middle part of my career, I took a role outside the statistics function as Technical Leader in the Strategy and Decision Science group. In this role, I was accountable for the summaries and evaluations of the company's drug research portfolio and the research on ways to optimize the portfolio. However, the new ideas and approaches that I helped develop were not readily adopted in the decision science group. In retrospect, I pushed too hard too soon. Although I was hired into that group with a mandate to explore new approaches, I failed to recognize the challenges of being a newcomer with new ideas in a long-standing, stable group. I would have been more successful if I had gradually introduced new ideas and been better able to couch those new ideas as an extension of existing ideas.

I made a similar mistake later in my career working within the statistics function at a different company. In this role, the major focus was the statistical education of a younger staff and to help implement statistical innovation. Although the statistics vice president (VP) supported these efforts, he was new to the company.

The day before the function-wide rollout of the statistical innovation group that I was to lead, a key member of the statistics leadership team sent a long email pushing back against the need for the group. Change is hard enough when everybody is on board. With key people working against change, the statistics innovation group was dead before it began.

I failed to recognize the culture of yes (see Chapter 5) wherein people appeared to agree to things in meetings but then disagreed outside the meetings. I again misjudged the appetite for change, and I pushed too hard for too radical an agenda that from the point of view of others stood apart from what had come before rather than building upon it.

3.5.4 Two Examples of Utilizing Scatterfocus

Although I have had many experiences in which Scatterfocus thinking contributed to creativity and innovation, the following example is perhaps the most notable. The team I was supporting was stuck on a problem. We were attempting to design a phase 4 clinical trial that would supplement the phase 3 data to clarify the optimum dosing regimen among the FDA-approved dosing regimens.

The specific aspects of the problem are not necessary. The key point was that we had been stuck for weeks in team discussions and meetings, with no prospect for consensus on a design for the trial. I had a one-week vacation scheduled in which the focus was wildlife photography. Although my vision impairment is severe, with powerful binoculars and telephoto lenses I can get some good pics – and I enjoy being in nature.

I left for this vacation frustrated by our trial design problem. But as the week progressed, with many quiet hours sitting in nature, interspersed by invigorating walks to move to new vantage points, stress and frustration melted, and I was not thinking about work. Figure 3.3 is from one of my wildlife photography trips. The Colorado mountains, sky, and wildlife provide a great opportunity for Scatterfocus thinking.

Upon returning to the office, my first meeting was with my Team's Medical Director to discuss the trial design problem.

FIGURE 3.3 Picture from a wildlife photography trip that fostered useful Scatterfocus thinking.

While I had been on vacation, other key members of the team met with experts in the research field relevant to our trial design problem to seek their advice. At the start of the meeting, the Medical Director reported no progress had been made through consulting with the experts.

When I was waiting in the airport for my flight home, which was delayed several hours, I grabbed my work notebook, which was my version of a common book where I kept thoughts and ideas. I drew out a design that somehow was just in my head. Without thinking about the problem for days, an idea was just there.

I shared this idea in my meeting with the Medical Director after hearing no progress had been made with the experts. We discussed the design, clarified a few points, tweaked it a bit, wrote out a list of strengths and limitations, and left the meeting confident that we had a solid proposal. At a team meeting later that day, consensus was reached to go forward with the design that had seemed to appear from nowhere while I was on vacation.

Of course, knowing the principles of Scatterfocus (Chapter 2) and creativity and innovation (Chapter 3), it is clear the solution only seemed to be a eureka moment. I had been thinking about it for weeks, storing thoughts and ideas in my common book. In working with the team on the design problem, I had collected all the dots. Going on vacation, relaxing, and minimizing the amount of phase-lock thinking fostered the necessary connections to be formed subconsciously.

While I was photographing wildlife in the beautiful scenery, with impressive sunrises and sunsets, my teammates were sitting in a stuffy hotel conference room rehashing the same old ideas getting nowhere. What a great lesson!

3.5.5 Playing Fetch with Maggie

Scatterfocus thinking does not require grand adventures, almost any relaxing scenario can work. From 2003 until 2018, Maggie, a yellow Labrador Retriever, was part of our family. Not surprisingly, Maggie loved to play fetch. We took her on walks or runs or played fetch every day.

The fetch sessions were good times for me to unwind after work, perhaps processing the day's activities or letting my mind wander as I stood in the park throwing the tennis ball for Maggie. Figure 3.4 shows the picture from a fetch session on a cold, quiet November morning.

I wish I had kept track of the many good ideas I had during those fetch sessions. After getting a promotion at work, my wife suggested that I should get Maggie a treat because she deserved most of the credit!

FIGURE 3.4   A good opportunity for Scatterfocus thinking – playing fetch with Maggie in the park on a frosty morning.

It often seemed difficult to find the time to fit the fetch sessions into the day. But in the long run, they paid dividends because I have many fond memories of Maggie and spending quiet hours in pretty places playing fetch, which led to good ideas. I found golden nuggets while playing fetch with Maggie!

CHAPTER 4

Communication

ABSTRACT

This chapter covers communication skills, with separate sections
for presentations and writing. Although some of us are inherently better presenters or writers, we can all improve. The section on presentation skills focuses first on major presentations
where substantial time to prepare, rehearse, and revise exists.
The principles for major presentations are then adapted to routine, on-the-job settings where preparation time is minimal and
little opportunity to rehearse and receive feedback exists. The
section on writing includes some of the basic elements of grammar, but also focuses on the ability to explain complex ideas in
simple terms, which is perhaps the most important aspect of
both verbal and written communication for statisticians. The
sections on presentation skills and writing each includes subsections with practical ideas on how to put the principles discussed
in the main body of the section into practice, with a focus on how
to use routine experiences on the job to improve our verbal and
written communication skills. This chapter concludes with some
real-world examples of the author's career.

DOI: 10.1201/9781003334286-5

4.1 INTRODUCTION

The importance of communication skills in career success is clear. Having insight, knowledge, or ideas is of little value if they cannot be communicated to others. Being good at written and oral communication requires practice. However, the following quote points to an important but often overlooked benefit of communication.

> It is not so much that I gave talks or wrote papers on things that I mastered; instead, I mastered the things I wrote and talked about. Presentations and writing were foundational parts of developing deep understanding and connecting the dots.

The Deep Work required to prepare good presentations and to write good papers fosters learning more about the topic. In addition, we think in our subconscious about the topic, thereby fostering the structured mind wondering, i.e., the Scatterfocus that can lead to new connections and new ideas. As discussed in Chapter 7, writing papers and giving presentations are a great way to continue learning, not just to tell others what we learned in the past.

As in creativity and innovation, some have inherently better communication skills than others, but everyone can improve with practice. Some shy away from situations that require effective communication because it is difficult. Some may feel their communication skills are inadequate, and failing in a presentation or communication is a public embarrassment. We think that because we are not currently a great speaker, or because we lack the brevity and wit of Abraham Lincoln, or the eloquence and imagery of Martin Luther King, we have nothing useful to say.

If you are stuck in this situation, consider Stephen Hawking, one of the most sought-after speakers in science through the last part of the 20th and early 21st centuries. Was Hawking a good speaker – Yes, but he could not speak! A neurodegenerative disease, amyotrophic lateral sclerosis, confined Hawking to a wheelchair. He could not move and gesture, he could not raise and lower his voice for effect, he had no voice at all. Earlier in the progression of his illness, Hawking used his finger to control a computer and voice synthesizer to choose his words. After he lost use of his hands, he twitched a cheek muscle to cue the cursor and choose a word or phrase.

Listening to Stephen Hawking was a slow and difficult process, yet his audiences were riveted. How was Hawking such an influential speaker under such difficult circumstances? He proved that if you have something important to say, there are many ways to get the idea across. He proved that the most important ingredient of communication is having something useful to convey (Figure 4.1).

FIGURE 4.1 Stephen Hawking giving a presentation.

Although there are many ways to get a point across, a few fundamental principles are key. This chapter highlights some of those key principles in oral and written communication. The section on presentations is based primarily on *The Official TED Guide to Public Speaking* (Anderson, 2016). This information is most relevant to major presentations for which considerable time to prepare exists. An additional subsection is devoted to adapting these principles to everyday settings where we communicate important information without days to prepare and practice. The section on writing is a brief overview of important concepts. This chapter covers foundational knowledge, but by itself may have minimal impact on communication skills. We must practice; therefore, an important part of developing good communication skills is to be in a position where these skills can be practiced regularly. Both the presentation and writing subsections include summaries on putting the principles into practice.

4.2 PRESENTATION SKILLS

4.2.1 Major Presentations

The Official TED Guide to Public Speaking (Anderson, 2016) is geared to 'major' presentations. For statisticians, major presentations might be a talk at an invited session at a major statistical or clinical conference, a presentation at an FDA Advisory Committee meeting, or a job interview. The common denominator is that we have time to prepare, revise, and rehearse the presentation. The following are some tips on giving good presentations; these are tips, not rules because good talks come from speakers who use their own style. Do not try to give a talk like someone you admire. It will not work. Nevertheless, you can incorporate successful elements from others.

Anderson (2016) suggests considering the four distinct parts in the process of developing a good presentation listed in Table 4.1.

TABLE 4.1   The Four Parts in Developing a Presentation

Start with the right foundations
Develop your ideas
Prepare
Delivery time

4.2.1.1 Start with the Right Foundation

Anderson (2016) suggests the most important thing in a great talk is to start with the right foundation, to have an idea you care about that is worth sharing. This does not have to be an invention or something ground-breaking. Anything that changes perspective, a how-to tip, or a review of key issues in a topic area. Do not underestimate the value of your work, learning, or insights.

Commitment is also important because giving a good talk is hard work (Anderson, 2016). For a major presentation, you should commit to multiple rounds of refinement. You may need to do research even if you are knowledgeable on the topic. This background work is essential in building your idea effectively and is best done early – typically before you begin drafting slides. The tendency is to start making slides as soon as possible. This makes us feel good, like we are accomplishing something. However, this progress comes at the cost of not having fully developed the idea, a shortcoming that will hinder the development of an effective talk throughout the process.

Through lines are often the cornerstones of great talks (Anderson, 2016). A through line is the core theme that connects components into a coherent picture. Great movies, plays, or novels usually have a through line. Think of a through line as the path the audience follows with the speaker to a destination. Effective through lines do more than state the topic or a key conclusion, they also give the audience something to think about. Good through lines are typically short, because short is simple, and simple is easier to believe or accept. Here are some examples:

- More choice makes us less happy

- Practice makes permanent

- On-the-job learning is key to individual and organization success

- I have a dream

The general approach to preparation is to find your through line and then build your talk around it. The goal is to help the audience to understand that line thoroughly without leaps in understanding or mental gaps. This focus can also help keep talks within the time limit by making clear what is essential versus extraneous information.

Most speakers complain they have too little time for what they want to say. The trouble is not time, it is how much we want to say (Anderson, 2016). When asked to speak on a topic, the tendency is to start by putting down everything we know and then try to compress that information into the time allotted. This approach almost guarantees a time crunch and a boring talk. Instead of starting with everything, start with a through line and only hang on what is essential. More specifically, define one core idea. Explain why it matters. Support key points with facts, stories, or examples. These key points are all linked to the core idea.

Some commonly used structures include (Anderson, 2016):

- What – So what – Now what

- Intro – Context – Key concepts – Implications – Conclusion

- Intro – Methods – Results – Discussion – Conclusion

4.2.1.2 Building Your Idea
Once you have a through line, start building the elements to hang on it. Five core tools for building elements of a talk are (Anderson, 2016):

- Connection
- Narration
- Explanation
- Persuasion
- Revelation

Connection is a tool to plant seeds in the minds of the audience. When we build a bond with the audience, they are more open to what we have to say. Approaches to establishing a connection include eye contact, laughter, an anecdote, a funny remark to cover a glitch, satire, visuals, and showing vulnerability. One of the factors contributing to Stephen Hawking's ability as a speaker was showing vulnerability. By putting his disability on display for all to see, by the audience seeing how hard Hawking had to work, a connection was formed, and Hawking was able to plant his ideas in the minds of others (Anderson, 2016).

Narration can be used to make a core idea more compelling, to add interest to the story. Four types of narration are (Anderson, 2016):

- Using a character your audience can relate to or care about
- Build tension using mystery, drama, or danger
- Use the right amount of detail. This element of narration is particularly important in scientific presentations. Include only what is essential. Know your audience. The level and types of detail useful to a statistical audience will differ from what is useful to a non-statistical audience.
- End with a proper resolution that is surprising, insightful, funny, or moving, and tie in with your core message. Usually, in a scientific talk, the resolution will be insightful.

Explanation is a technique to build understanding by layering information in a hierarchy, with each layer constructed on a previous layer. To explain something complex or technical, start with what the audience already knows and builds from there. Do not assume the audience is familiar with your topic and technical jargon. Consider using metaphors to draw parallels between what you are explaining and something that the audience already understands. For example, the following metaphor may help an audience understand how the brain works. 'The pre-frontal cortex of the brain works like a flight simulator that lets us imagine things.' As new layers of information and understanding are built, short examples can be used to apply the idea, thereby reinforcing understanding (Anderson, 2016).

Persuasion is how we convince others. Sometimes, to explain a new idea, we must first tear down existing concepts that are wrong. We can prime or nudge others in a certain direction by using metaphors to make an idea more plausible (Anderson, 2016). Priming is covered in more detail in Chapter 5 on critical thinking and making decisions under uncertainty. Once someone is primed, it is easier to present arguments using reasoning.

Much of effective persuasion in science is in ordering the arguments. Consider parsing your argument into a series of if-then statements. Start with something well known and accepted (or was proven earlier in the talk). By building from what is known the audience begins by agreeing with you, which can make it easier for them to continue agreeing with you on later points. Also, breaking down the argument into components can make the position easier to understand, and what is easier to understand is also easier to believe.

Revelation, or showing an idea, is the most direct way to convey an idea. For example, when speaking about a new approach, technology, process, or invention, a demonstration is one of the most effective ways to show how it works (Anderson, 2016).

4.2.1.3 Preparing the Presentation

Preparing a talk can be broken down into the following components (Anderson, 2016):

- Visuals

- Scripting

- Rehearsing

- Opening and closing

Visuals (photos, illustrations, graphs, etc.) are not always essential and bad visuals can hinder understanding. A good table may be better than a bad graph. Use visuals only if necessary to make the talk more effective (Anderson, 2016). Too many visuals distract and reduce the impact of essential visuals. However, when used properly, visuals can reveal something that is hard to describe in words; a picture paints 1,000 words. Visuals can also enhance demonstrations and build curiosity or understanding.

Expert advice on whether a talk should be scripted is to let speakers do whatever works best for them. Scripted talks can be efficient because the speaker will not wander 'off script.' However, scripted talks must be delivered properly to avoid losing passion and connection. We should not read a scripted talk. Three basic strategies to build a scripted talk are (Anderson, 2016):

- Know the talk so well that it does not seem scripted. This takes a lot of practice, until you know the material so well that you do not have to read it from a script and instead you can 'talk' to the audience

- Refer to the script from a lectern, screen, or confidence monitor, but look up during every sentence to make eye contact with the audience so you are still speaking to them, not just reading

- Reduce the script to bullet points and express each point in your own words. Unscripted talks tend to sound fresh because the speaker is thinking aloud and searching for the words at the moment. However, you still need to prepare for unscripted talks

Regardless of the approach to developing a major presentation, rehearsing is a must (Anderson, 2016). The more realistic the rehearsal situation, the more useful the rehearsal. Start by rehearsing in whatever manner is comfortable. Then rehearse in front of a mirror, then in front of friends, family, or a peer group. For example, if you have a major presentation at a conference in early August, prepare early and consider giving the talk as a seminar to your function or work group a month or more before the conference. We benefit from honest feedback on our rehearsals. Knowing 'that was good' will not help us improve – and the goal of rehearsal is to improve.

Even when not scripting the entire talk, it is useful to memorize the opening and closing lines. This ensures the clearest communication of the most impactful parts of the presentation. Strategies to create a strong opening include (Anderson, 2016):

- Use drama to make people curious about your talk. 'In the next 18 minutes, as I give this talk, four Americans will die from the food they eat'

- Tell a vivid story or a humorous line

- Spark curiosity by asking a surprising question. 'How did a 10-year-old with only $300 get an entire town to change their future?'

- Use an impactful visual or object to capture attention, and if something surprising is revealed, it creates an even bigger impact. We can also use openings like: 'The video I'm about to show changed my life' or 'I'm going to show you something that may seem impossible at first glance'

- Use a teaser that gives a sense of where you are at and where you are going. But do not give away the punchline or the big reveal right at the start

A strong closing is important because the ending shapes how people remember the entire experience. Ways to close strong include (Anderson, 2016):

- Pull back to show the big picture and the implications and possibilities of what you have just shared

- Give a specific call to action and nudge the audience to act on the powerful idea you've just shared

- Make a personal commitment on what you are going to do

- Create a vision of what the future might be

- Ask a surprising question that makes people think about what you have said

- Use narrative symmetry by linking back to something you have said in the beginning

4.2.2 Putting the Principles into Practice

Here are some additional tips on putting the principles discussed in the previous section into practice. Tips for getting the right foundation include (Anderson, 2016):

- Great public speaking does not come from following a fixed formula or trying to be someone else. Be authentic and speak about a topic you care about

- The best way to communicate an idea is to recreate the idea in the audience's minds. This requires beginning where the audience is and bringing them on a journey with us

- Use your audience's language, including examples and terms to which they can relate. For example, most people can imagine an elephant, but not a *Loxodonta* (an elephant's scientific name). Trying to impress an audience with esoteric knowledge is a good way to lose the audience, fast

- Pay attention to your choice of words, body language, and tone of voice. They are all vital to your talk

Things you should *NOT* do in getting the right foundation include (Anderson, 2016):

- Make a sales pitch: Your role as a speaker is to give to the audience, not take from them

- Ramble pointlessly: Rambling without a clear message or takeaway is disrespectful of the audience's time

- Use company talk: People who do not work in your organization are not interested in your mission, products, or milestones. However, they would want to hear about the ideas behind your work, what you learned, and how they can apply those insights to their work

- Put on an act or show merely for show. Many speakers try too hard to be liked by the audience

Tips on connecting with the audience include (Anderson, 2016):

- Shelf your ego: It is hard to realize when we are being egoistic. Therefore, it is important to check that we are not including a line(s) to name-drop or to boast about our achievements. Test the talk on someone you trust to get feedback

- Tell stories about our personal experiences or people close to us. These can be used in any part of the talk

- Eye contact and an occasional smile or appropriate expression is a good way to connect with the audience

Tips on using humor include (Anderson, 2016):

- Humor is a skilled art, and not everyone can do it. However, humor is just one of many tools. If we are not good at using one tool, we can use another

- Ineffective humor is worse than no humor at all

- Self-deprecation, done properly, can be engaging, show vulnerability, and make connections

For many statisticians, explanation is the most important skill to improve upon. Tips on explanation include (Anderson, 2016):

- Helping someone to understand a concept fundamentally changes their mental model and creates a lasting impact

- To give someone a new idea or important information, we must be able to explain it. Some are naturally better at explanation than others, but it is a skill that can be improved. If you have something important to say and the audience understands your point, it will not matter that much if it is read verbatim off a cue card or delivered with soaring rhetoric

- Explain in layers or a stepwise progression. Get feedback on how much detail is needed. We tend to overcomplicate and overexplain

- Do not tell the audience everything you know, tell them what they need to know and why it is important to them

Tips on persuasion include (Anderson, 2016):

- Priming makes an audience more persuadable

- Couch your work as a detective story: Present a mystery, the possible solutions, then eliminate the options one by one until there's only one logical conclusion. Most scientists will react better by being shown the trail of facts and findings so they can come to their own conclusion rather than being told what or how to think

- Go beyond logic. Statisticians rely more heavily than most on logic. Nevertheless, humor, anecdotes, vivid examples, third-party validation, and impactful visuals enhance the logic of arguments to make them more meaningful and energizing. This is especially true when we present to non-statistical audiences

Tips on delivering the talk include (Anderson, 2016):

- Public speaking makes everyone nervous

- Nervousness fuels early preparation in good speakers

- A few pointers on managing anxiety during the presentation include:

  - Breathe deeply and exhale slowly in the moments before going on stage

  - Avoid an empty stomach or eating too much

  - Drink some water

  - Identify a few friendly faces in the audience and focus on them

  - Have a backup plan for common problems like a technology glitch

  - Remind yourself that what you are sharing matters and focus on sharing it

The guidance in this section is geared toward major presentations, such as a talk at a scientific conference which is scheduled months in advance, thereby affording an opportunity to develop and refine the presentation. However, most of the presentations we give are not major presentations; they are routine presentations given as a regular part of our job with minimal time to prepare. Nevertheless, these routine presentations are opportunities to practice the principles noted in previous sections, albeit in a more condensed manner.

In preparing for major or routine presentations in scientific settings, remember that a clear explanation of often complex topics is key. Yes, it will take more time to draft and revise a presentation than to merely put together some slides. It may seem that the extra time spent on a routine presentation is hard to justify. However, the benefit far outweighs the cost.

First, your presentation skills will improve, and it will therefore take less time and effort as time goes on. Second, your presentations will be clearer and easier to understand. If your points are clearer, they are more likely to be accepted, and the audience will justifiably associate your clear presentations with clear thinking; that is, you will be seen as a smarter and more effective statistician. Moreover, the practice gained from improving routine presentations will directly translate to more effective major presentations.

In developing the first draft of a presentation, whether major or routine, careless preparation will result in a greater need for revision. However, too much effort spent refining content at the initial stage may inhibit assembling the most useful content. No matter how careful we are with an initial draft, much work will remain to be done. Hence, do not try to make the first draft perfect. Focus on getting the main ideas down, knowing revisions will be necessary.

Feedback from others is especially useful in assessing whether the original content assembled for a talk is appropriate for the circumstance. Feedback can also refine the presentation to help ensure explanations are clear.

FIGURE 4.2 The author giving a presentation at the 2022 Joint Statistical Meeting.

Whether the presentation is major or routine, be sure you can stick to the time constraints. No matter how well-refined and clear your presentation, if the material does not fit into the allotted time the presentation will be ineffective (Figure 4.2).

4.3 WRITING

4.3.1 Introduction

Many parallels exist between good presentations and good writing. Like presentations, good writing could be a book by itself; hence, only the most pertinent details are covered here. And the most important aspect of good writing, as with presentations, is to practice, and whenever feasible seek feedback and revise accordingly.

Abraham Lincoln's Gettysburg Address is an instructive example. The address was, as the name implies, a speech. However, in that era, the text of speeches was important because few people could travel to see speeches live, and of course TV and video did not exist. The only way to ensure a wide audience had access to a speech was through reprinted versions in newspapers and pamphlets.

Lore has it that Lincoln wrote the text for the Gettysburg Address at the last minute, aboard the train on his way to

Gettysburg. This fed into the perception that Lincoln's ability as a great communicator was inherent and natural. But the story is not true. Yes, Lincoln was a gifted communicator, but that was not the product of inherent ability alone. He honed his skills through years of practice and feedback, and improved his written and oral communication through revision.

Five known copies of the speech in Lincoln's handwriting exist, each with different wording and punctuation. These five copies were named for the people who first received them. The clear and even script in each copy is consistent with writing on a hard, steady surface rather than the bumpy trains of that era. The purpose of sharing early drafts was to solicit feedback from trusted advisors. The lesson for us is that while some writers are better than others, all good writing requires revision, and no one becomes a good writer without practice. Being a good writer might be glamorous but becoming a good writer and good writing itself is not glamorous – it is hard work, even for the most gifted (Figure 4.3).

4.3.2 Explanation

As with presentations, the most important aspect of effective writing for statisticians is the ability to explain. The 'language' of mathematics provides the precision statisticians need to communicate with each other, to explain the nuances of statistical methods and analytic models. However, when statisticians communicate with non-statisticians, mathematical language is often not effective because non-statisticians lack the necessary background. Often, statisticians do not need to communicate with non-statisticians with the same degree of precision and nuance as when communicating with other statisticians. The goal in communicating with non-statisticians is often to convey key concepts and inferences rather than statistical detail and nuance.

Fundamental to effective communication with non-statisticians is an appreciation for what others need to know, rather

FIGURE 4.3 The Gettysburg Address inscribed on the South Chamber Wall in the Lincoln Memorial.

than what we want to say, or saying everything we know about the topic.

Consider the following three versions of a paragraph that is intended to convince readers that they should read the book *The Elements of Style* (Strunk & White, 1999) to improve their writing.

Version 1:

The most useful, relevant, and insightful instruction I have ever received on writing was via the excellent, well-known, timeless classic practical guide to writing by Strunk and White, *The Elements of Style*, the first vestiges of which appeared by way of Strunk's prodigious pen in 1918 and was published by Harcourt publishing in 1920, but was greatly expanded upon by E. B. White (Charlotte's Web) with the first edition of the expanded version of *The Elements of Style* published by Macmillan in 1959, and has since been named by the *Time* magazine as one of the 100 best and most influential books written in English since 1923.

Version 2:

The *Elements of Style* is a classic book that has improved the writing of countless authors for generations. The book includes eight 'elementary rules of usage,' ten 'elementary principles of composition,' 'a few matters of form,' a list of 49 'words and expressions commonly misused,' and a list of 57 'words often misspelled.'

Version 3:

The book, *The Elements of Style*, will improve your writing.

Which version is best? The three versions have 106, 53, and 10 words, respectively. The common, commercially available software used to prepare this book did not flag any grammar, usage, or spelling errors in any of the versions.

Version 1 has more detail than the other versions and is a first-person account from an author who used the book. Therefore, we might expect Version 1 to be the most persuasive, but it is not. Version 1 is long and convoluted, with many details that are ancillary to the goal of convincing readers on the merits of the book. Version 1 also contains unneeded adjectives and adverbs, and convoluted punctuation. The author appears to be trying to tell readers everything he/she knows about the book rather than focusing on why the book would be useful to the reader.

Version 2 is shorter and more direct than Version 1. The first sentence is the main point and specific, objective details about the book's content are enumerated in the second sentence. However, readers may lack context about the details, the numbers, in the second sentence, which could lead to misinterpretation. Is the book long or short; is it detailed and technical or practical and to the point?

Version 3 is a short and simple declarative statement without supporting details. However, without details, the version may be unconvincing.

The strengths and limitations of these three versions lead to a fourth option that combines the simple declarative statement as in Version 3 and the first sentence of Version 2, with a shorter, more general summary of content than that used in Version 2.

> *The Elements of Style* is an award-winning book by Strunk and White that has improved the writing of countless authors through its short, practical approach.

This example illustrates some fundamental guidelines for good writing.

- Shorter is usually better

- Do not leave out essential details

- Do not include unneeded details

Understanding what details are essential versus unneeded in explaining something, especially a complex statistical idea, requires knowledge of the topic and the audience. Achieving the proper balance in detail for an explanation usually requires revision. Mastering key aspects of grammar will not help much in achieving the proper amount of detail in explanations. However, effective grammar, like presentation skills, will help to convey explanations more clearly.

4.3.3 Some Elements of Style and Grammar

The many rules of grammar can make writing intimidating. However, mastering a few key points will lead to effective communication. These points include voice, person, tense, and usage of pronouns, adjectives, and adverbs. This section provides examples of these key points.

4.3.3.1 Voice

The two main types of voice in writing are active voice and passive voice. In passive voice writing, the subject receives an action. In active voice writing, the subject performs an action. Passive voice often leads to unclear, indirect, and wordy sentences, whereas active voice creates clearer, more concise sentences.

The following examples illustrate active and passive voice writing.

Active: Smith et al. studied the effects of social support on recovery from PTSD.

Passive: The effects of social support on recovery from PTSD were studied by Smith et al.

Active: Craig ate fish and chips at dinner.

Passive: At dinner, fish and chips were eaten by Craig.

Active: We are going to write a protocol.

Passive: A protocol is going to be written by us.

Active: Sarah programmed the sensitivity analyses.

Passive: The sensitivity analyses were programmed by Sarah.

4.3.3.2 Tense

Although no hard and fast rule prevents changing tense within a sentence or a paragraph, such changes should be used only when necessary. Grammarians define 12 tenses in writing. However, three tenses make up most of the tense verbs used in academic writing: Present simple, past simple, and present perfect. These tenses can be used both in passive and active voice. The following are common functions that these three tenses have in academic writing, along with examples.

4.3.3.2.1 Present Simple Tense
Functions:

- Framing the paper in the introduction, such as stating what is known about the topic, and in conclusions to state what is now known

- Point out the aim of the current paper

- Refer to findings from previous studies without mentioning the author's name

- Refer to tables or figures

Examples:

- Statisticians <u>are</u> aware of the need to assess the robustness of inferences to departures from assumptions.

- The primary objective of this investigation <u>is</u> to assess the effectiveness of fluoxetine versus placebo.

- Diet and exercise used in combination <u>is</u> more effective than either used alone.

- Results from the primary analysis <u>are</u> summarized in Table 4.1.

4.3.3.2.2 Past Simple Tense
Past simple tense is used to refer to actions completed in the past.
Functions:

- Report the findings of a previous study

- Describe the data and methods used in the study

- Report the results of the current study

Examples:

- Smith et al. reported that the variability in placebo response <u>was</u> a major factor in the high rate of failed studies in major depressive disorder.

- The primary analysis <u>was</u> based on a mixed-effect model.

- The mean change from baseline to endpoint for placebo <u>was</u> 6.1.

4.3.3.2.3 Present Perfect Tense

Present perfect tense is used when referring to previous research, and since it is a present tense, it indicates that the findings are relevant today.

Functions

- Introduce a new topic, or a new report

- General summaries of previous research

Examples:

- A large body of research <u>has shown</u> that missing data is a source of bias in estimates of treatment effects.

- Several meta-analyses <u>have found</u> that the fraction of patients randomized to placebo influences the magnitude of placebo response.

In everyday communication such as emails, the future tense is common. In these situations, the future tense can communicate upcoming actions or events.

Examples:

- I will finish the protocol by Tuesday.

- A meeting will be scheduled to discuss the statistical analysis plan.

4.3.3.3 Person

First, second, and third person writing are ways of describing points of view. First person is the I/we perspective. Second person is the perspective of another specific person. Third person is the he/she/it/they perspective. Any of these points of view can be used with active or passive voice, and with any of the common tenses.

4.3.3.3.1 First Person

First person is used to talk about ourselves, our opinions, and the things that happen to us. The biggest clue that a sentence is written in the first person is the use of first-person pronouns. We, us, our, and ourselves are all first-person pronouns. Specifically, they are plural first-person pronouns. Singular first-person pronouns include I, me, my, mine, and myself.

Many stories and novels are written in first person. In this kind of narrative, readers are inside a character's head, watching the story unfold through that character's eyes. First-person writing is uncommon in scientific research but common in everyday communication.

Examples:

- I found the statistical methods section difficult to understand.
- I fit a mixed-effect model.

4.3.3.3.2 Second Person

The second-person point of view belongs to another person(s), or the persons(s) being addressed in the sentence. The biggest indicator of second-person writing is the use of second-person pronouns: You, your, yours, yourself, yourselves. Second-person writing is common in conversation, but uncommon in stories, novels, and scientific papers.

Examples:

- You can wait in here and make yourself at home.

- You can fit a mixed-effects model.

- You should be proud of yourselves for finishing this enormous project!

4.3.3.3.3 Third Person

The third-person point of view belongs to the person (or people) being talked about. The third-person pronouns include he, him, his, himself, she, her, hers, herself, it, its, itself, they, them, their, theirs, and themselves.

Examples:

- She used a mixed-effects model in her research.

- They found the statistical methods section difficult to understand.

Pronouns do not always reveal whose perspective is represented in a sentence because not all sentences include pronouns, especially in the third person: Consider the sentence:

Mike always fits mixed-effects models.

Mike is not the writer talking about his own work; hence the sentence is not first person. The writer is not talking to Mike, so that eliminates second person. Hence, the sentence must be third person.

Many stories and novels are written in third person. In this type of story, a disembodied narrator describes what the characters do and what happens to them. We do not see directly through a character's eyes as in a first-person narrative, but often the narrator describes the main character's thoughts and feelings about what is going on.

In scientific writing, personal perspectives should usually be avoided, which means that first-person writing is not usually appropriate in these settings. For example, do not write

> I think the results of the EMPOWER study indicate more research is warranted.

Instead, adopt a third-person perspective using citations whenever possible, such as:

> Results of the EMPOWER study suggest more research is warranted (Smith et al).

When writing statistical methods sections about our research, it is tempting to use first person because we are the ones who did the work. However, this leads to frequent use of I and we, which can become irritating to readers. For example:

> We randomized patients in a 1:1 ratio and we fit a mixed-effects model for the primary analysis.

Should be written as:

> Patients were randomized in a 1:1 ratio and a mixed-effects model was used for the primary analysis.

The context of a scientific paper is such that readers already know it is the authors of the paper whose perspective is represented and who did the work, unless citation to others indicates otherwise.

4.3.3.4 Pronouns
Pronouns are used to replace nouns. Different types of pronouns are used to indicate person, number, gender, and case. Pronoun

types include personal, interrogative, indefinite, demonstrative, and reflexive. These pronoun types are not discussed in detail here; however, examples of how vague pronouns can be an impediment to clear writing are discussed.

Consider the following three examples:

Example 1: Craig ran the 1,500-meter semi-final with Leamon two hours after he ran the 5,000-meter final. It was the hardest race of his life.

It is not clear who ran the 5,000-meter final, nor is it clear which race was the hardest of whose life. A clearer way to communicate this information would be:

Craig and Leamon ran the 1,500-meter semi-final, which was only two hours after Craig ran the hardest race of his life in the 5,000-meter final.

Example 2: The room contained a chair, lamp, and desk. It was 20 feet long.

It is not clear whether the room or desk was 20 feet long. Clearer wording would be:

The room was 20 feet long and contained a chair, lamp, and desk.

Example 3: Sally worked at a large pharmaceutical company before joining a CRO. That was a wise career decision.

It is not clear whether the wise decision was joining the CRO or first working in large Pharma. Clearer wording would be:

It was wise for Sally to work in large Pharma before joining a CRO.

4.3.3.5 Adjectives and Adverbs

Adjectives modify nouns; that is, adjectives provide more information or context about the noun. In the following examples, the adjective is underlined.

- Jack lives in a <u>beautiful</u> house.

- He writes <u>meaningless</u> emails.

- Ramona is an <u>adorable</u> baby.

Using adjectives effectively in scientific writing can be tricky because adjectives are easy to overuse, stemming from well-intentioned efforts to be detailed. However, too many adjectives can be distracting, or worse yet, especially in scientific writing, foster the belief that the writer is exaggerating. Adjectives may also be interpreted by the reader as the subjective opinion of the author, thereby fostering unnecessary disagreement in scientific writing.

Adverbs are typically used to modify a verb, adjective, or another adverb to convey manner, place, time, frequency, degree, level of certainty, etc. Adverbs answer questions such as how, in what way? when? where? and to what extent? In the following examples, the adverb is underlined.

- He swims <u>well</u>.

- Alex ran <u>quickly</u>.

- Samantha spoke <u>quietly</u>.

- James coughed <u>loudly</u> to attract her attention.

- He played the guitar <u>beautifully</u>.

In scientific writing, adverbs can be used to demonstrate a point. Reference can be made to a scientist who 'regularly' expresses a viewpoint or bias by using an adverb to show the extent of their credibility as a source. Or a source cited in a paper can be said to have 'expertly' stated a case, or to have 'eloquently' made their point. In these cases, the adverbs add plausibility or credibility to why they were cited.

However, as with adjectives, the value of adverbs in scientific writing relies on their being used infrequently and that a better alternative does not exist. Moreover, some adverbs should be avoided in scientific writing. For example, intensive adverbs such as very, truly, really, actually, and extremely often weaken instead of intensifying the words. Consider the following three sentence fragments:

- ... a really important sensitivity analysis ...

- ... an important sensitivity analysis ...

- ... a crucial sensitivity analysis ...

A 'really important' sensitivity analysis does not sound professional and is not more consequential than an 'important' or 'crucial' one.

In the following example, the initial wording uses adjectives and adverbs. The second version replaces adjectives and adverbs with objective facts that are more convincing to a scientific audience.

Version 1:

Dr Smith is a truly great statistician with extensive expertise in group sequential methods.

Version 2:

Dr Smith is a Fellow of the American Statistical Association who won the Royal Statistical Society's award for excellence in the pharmaceutical industry based on her contributions to group sequential methods.

This example illustrates an important general point in scientific writing/presentations and general communication: It is often better to show the reader (audience) than to tell the reader. That is, show the reader/audience the facts so that they can come to the same conclusion as you rather than telling the audience your opinion and hoping they agree with you.

4.3.4 Putting the Principles into Practice

How many emails do you send in a day, week, or month? Each email provides an opportunity to practice good writing. Yes, it will take more time to draft and revise an email than to type it out and click send, but the benefit far outweighs the cost. First, your writing will improve, and it will therefore take less time and effort as time goes on. Second, your emails will be clearer and easier to understand. If your points are clearer, they are more likely to be accepted, and readers will justifiably associate your clear writing with clear thinking; that is, you will be seen as a smarter and more effective statistician. Moreover, the practice gained from improving emails will directly translate to more effective scientific writing in manuscripts, protocols, statistical analysis plans, etc.

In writing the first draft of a paper, email, or letter, careless writing will result in a greater need for revision. However, too much effort spent on grammar and editing in the initial draft can interfere with creativity. No matter how careful we are with an initial draft, much work will remain to be done. Hence, do not try to make the first draft perfect. Focus on getting the main ideas down as best you can, knowing revisions will be necessary.

The Elements of Style (Strunk and White, 1999) has been used in several examples in this chapter. Reading that book and putting its principles into practice is a time-efficient way to improve your writing. In addition, get feedback on your writing. Find a writing coach or mentor at your workplace and/or outside of work. Use feedback from your coach/mentor to improve your writing.

A simple approach to proofreading includes doing searches for commonly used pronouns to verify whether they are clear. Searches can be done for adjectives and adverbs to assess whether alternative wording would be more appropriate. In everyday writing, adjectives and adverbs can be used more often than in scientific writing, but it is always a good idea to confirm their usage.

As a last tip, use functions of common word-processing software to improve your writing. Use the reading function to read the paper or email it back to you. This will often catch mistakes that otherwise would get overlooked due to familiarity with the writing. It will also catch misspelled words that are not flagged by the software. For example, consider the following sentence.

Mixed-effects models are often used in the analysis of clinical trail data.

Having software read the passage will allow you to catch the typo *trail* which should be *trial*.

Many commercially available word-processing software packages also have features to perform the searches mentioned above, along with other grammar checks. These features are useful and should be included as part of the proofreading process. However, do not rely on these standard screens to catch all the opportunities for improving the writing in your document.

4.4 REAL-WORLD EXAMPLES

4.4.1 Presentations

The following are some examples of major presentations that I have given in which preparations were extensive, as outlined in the *TED Guide to Public Speaking* (Anderson, 2016).

In 2003, I was a presenter at an invited session at the Joint Statistical Meetings (JSM). The session was on accounting for missing data in clinical trials and included as speakers and discussants two world-renowned academics and one of the highest-ranking statisticians at FDA – and me. I did not have anywhere near the stature of the other participants. The session was in one of the largest ballrooms in the conference hall and drew a standing-room-only crowd. Not surprisingly, I was nervous.

Knowing the potential importance of the session, I completed my research several months before the conference. I prepared my talk soon thereafter and first practiced on my own. I then gave the talk to a small group of statisticians at a weekly forum in my workgroup. A few weeks later, I gave the presentation to a larger audience at a function-wide seminar at my company. At each presentation, I solicited general feedback, but also designated individuals to provide detailed feedback. By the time JSM rolled around, I was able to deal with the anxiety of the moment knowing I had thoroughly prepared. My opening and closing statements, along with a few key explanations, were scripted and memorized. Once I got past the opening, I was calm and the talk went well.

For a job interview early in my career, I gave a talk in which I prepared much as described above for the JSM talk. However, later in my career when I was interviewing for a job with a smaller group at a company where I already knew many of the key people, my preparations, while again diligent, were different. In the latter instance, I (correctly) anticipated a more conversational atmosphere, with more discussion and less time devoted to cranking through prepared remarks. In some ways, the discussion format can be harder to prepare for because you cannot anticipate every direction the discussion may go. When giving a formal presentation the speaker is in control, at least until questions at the end.

I have given as many as 25 presentations in one year. It is not possible, nor is it necessary, to prepare as extensively as described in the TED guide for every talk. With less time to prepare, we rely more on our general knowledge of the topic and cannot memorize much. In these instances, I tend to put key ideas or conclusions in a different color font, which serves two purposes. The coloring is my cue to explicitly cover this point, and it also draws the audience's attention to it.

With less time to prepare and memorize, I inevitably tend to put more text on each slide, which is a slippery slope. Too much text and one may as well be reading from a teleprompter, which will be boring for the audience. Therefore, I tend to use sentence

decide. At the organizational level, we must understand how structure, systems, and culture shape the information and inferences used in decisions (Roberto, 2013).

The next section of this chapter examines individual factors that influence decision-making. Group and organizational factors that influence decision-making are covered in Chapter 6, which is the first chapter in Part II – Working with Others.

5.2 INDIVIDUAL FACTORS IN CRITICAL THINKING AND DECISION-MAKING

5.2.1 How We Think

In *Thinking Fast and Slow* (Kahneman, 2011), Nobel Laureate Daniel Kahneman summarizes 40 years of research, much done with his primary collaborator Amos Tversky, on how we make decisions and think about problems. This work demonstrated that we often assume things automatically – without giving them careful thought. Kahneman calls those assumptions heuristics, and he provides extensive research results and examples of how heuristics can help and hinder our decision-making.

A compelling aspect of the book is Kahneman's conceptual model of how the brain works in making decisions. Kahneman introduces two conceptual characters. These characters are called System 1 and System 2. System 1 thinks fast, and System 2 thinks slow. Kahneman is quick to note that the goal is not to understand which system is best but to understand when System 1 is best, when System 2 is best, and how we can improve our judgment. An illustration of System 1 and System 2 thinking is provided in Figure 5.1.

System 1 operates automatically, intuitively, involuntarily, and effortlessly, such as when we drive, read facial expressions, or recall our age. System 1 thinking is therefore efficient and allows us to make thousands of decisions without bogging down our brains. System 2 requires slowing down, deliberating, reasoning, computing, focusing, concentrating, considering other data, and not jumping to quick conclusions. Examples of System 2 thinking

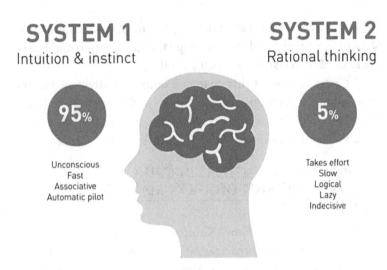

FIGURE 5.1   An illustration of System 1 and System 2 thinking.

include solving math problems, making investment decisions, writing computer code, etc. (Kahneman, 2011). System 2 involves the Deep Work and Hyperfocus discussed in Chapter 2. System 1 does not require Deep Work because it operates on heuristics, but in complex settings, these heuristics may not be accurate, and hence System 1 can be wrong. System 2 requires effort to evaluate those heuristics. The overriding themes in *Thinking Fast and Slow* are that humans are not 'wired' for making good decisions under uncertainty, how to recognize situations in which mistakes are likely, and how to avoid mistakes when stakes are high (Kahneman, 2011).

Kahneman (2011) explains that System 1 suppresses ambiguity and doubt by constructing coherent stories from data scraps and filling in those gaps with automatic guesses and interpretations that fit an often easy to understand, and therefore easy to believe, narrative. System 2 is the inner skeptic, weighing those stories, doubting them, and suspending judgment. However, because System 2 requires hard work, it sometimes fails to do its job, allowing us to slide into certainty. We have a bias toward believing.

System 1 tends to replace a tough question with an easier one. For example, when asked if we are happy with our life, responses tend to reflect how happy we are now, which is easier to consider than the totality of our life (Kahneman, 2011). In a business setting, if asked the probability a project will succeed, which implies a difficult-to-determine number on a continuous scale from 0 to 1, the tendency is to replace that number with a binary, often qualitative, response such as 'I think this project has a fighting chance'; or, 'this project looks better than most.' This qualitative response may then be back-translated to a high probability of success, which may be an overestimate.

System 1 relies on intuition and matching, thereby rating relative merit through matching what may be dissimilar traits, such as a funny person is often considered to be smart. System 1 is prone to evaluate a decision without distinguishing which variables are most important. The basic and potentially inaccurate assessments from System 1 replace the hard work System 2 must do to make judgment (Kahneman, 2011).

Kahneman (2011) further explains that these mistakes are the result of cognitive biases. The following sections summarize some of these biases as described by Kahneman (2011).

5.2.2 Cognitive Biases

Historically, economic models depicted individuals as rational decision-makers. That is, individuals collect information, examine alternatives, and make decisions that maximize personal satisfaction (utility). However, in *Thinking Fast and Slow*, Kahneman (2011) shows that we do not always make decisions consistent with rational economic choice; in fact, we seldom make decisions based entirely on rational economic choice.

An important factor that can cause deviation from rational choice is cognitive limitations that lead to errors in judgment.

Humans are hard-wired through genetics and centuries of environmental stimuli to behave in certain ways, and those ways are not always consistent with rational decision-making. Psychologists describe these systematic mistakes as cognitive biases. Think of these as decision-making traps that we are likely to fall into unless we work to avoid them through an appropriate decision-making process. We are not biased because we lack knowledge. We are biased because we are human.

Cognitive biases affect experts and novices in all fields of work and life. Psychologists have demonstrated the existence of these biases in experimental and real-world settings. The following paragraphs summarize many of the key cognitive biases discussed in *Thinking Fast and Slow* (Kahneman, 2011) that can influence decisions.

5.2.2.1 Affect Bias

In the affect bias, mood influences thinking such that more favorable responses are likely when we are in a good mood. Kids have long recognized affect bias – that is why they ask parents for what they want when the parents are in a good mood.

5.2.2.2 Anchoring Bias

Anchoring bias is when an initial reference point distorts estimates. We begin at the reference point and adjust from there, even if the initial reference point is arbitrary. In one of their studies, Kahneman and Tversky asked study participants to guess the percentage of African nations that were United Nations members. They asked some if the percentage was more or less than 45% and others whether it was more or less than 65%. The former group estimated a lower percentage than the latter. The initial reference points served as anchors. This bias can affect a wide variety of real-world decisions. For example, 'planning for success' in drug development or any research project may be useful operationally, but it may also establish an overly optimistic view on the probability of success.

5.2.2.3 Availability Bias

When asked to estimate numbers such as the frequency of divorces in Hollywood or the number of deaths by plane crash, the ease with which we retrieve the information influences the answer. We are prone to give bigger answers to questions where the information is easier to retrieve. And answers are easier to retrieve when we have emotional or personal experience, especially recent experience. Someone who got mugged overestimates the frequency of muggings. Someone who just completed a series of chores at home overestimates the percentage of the housework they do. Someone who has never seen or experienced addiction underestimates the risk of addiction. News of a recent plane crash causes us to overestimate the risk of air travel.

5.2.2.4 Coherent Association

To make sense of the world, we create narratives and make associations between events, circumstances, and regular occurrences. The more an event fits into our stories, the more normal and believable it seems. Things that do not occur as expected take us by surprise. To fit those surprises into our mental models, we devise new stories or narratives that make the unexpected event or information fit. Sayings such as 'Everything happens for a reason' or 'That was so weird it can't be due to chance' are common examples of explaining the unexpected without clear evidence.

Abnormalities, anomalies, and incongruities in daily living are key developments of explanations that are coherent with our mental models. These explanations often involve (1) assuming intention, 'It was meant to happen,' (2) causality, 'They're homeless because they're lazy,' or (3) interpreting providence, 'There's a divine purpose in everything.' Given that these explanations are often not grounded in objective information, we risk making errors such as positing intention and agency where none exists, confusing causality with correlation, and making more out of coincidences than is warranted.

5.2.2.5 Cognitive Ease

Things that are easier to compute, more familiar, and easier to read are easier to believe than things that require hard thought, are novel, or are hard to see. If a statement is linked to other beliefs or preferences, or comes from a trusted or likable source, we feel a sense of cognitive ease and believability, even without direct evidence for the statement itself.

Because things that are familiar are easier to believe, teachers, advertisers, marketers, authoritarian tyrants, and cult leaders repeat their messages. If we hear something often enough, we tend to believe it, no matter the amount or quality of evidence to support the statement, idea, or belief.

5.2.2.6 Confirmation Bias

This is one of the most prevalent cognitive biases in science and business. The confirmation bias is the tendency to gather and focus on information that confirms existing views and to avoid or downplay information that disconfirms preexisting hypotheses or beliefs. Examples include focusing on the best result (endpoint from a trial, best trial in a development plan) when assessing the overall merit of a new drug or focusing on the best result from a series of focus group assessments of a new product.

5.2.2.7 Conjunction Fallacy

The conjunction fallacy causes us to choose what we consider a plausible story over one that is in truth more probable. In a famous experiment, Kahneman and Tversky provided research subjects with the following details on a fictitious character, Linda. In the 1970s, when this research was conducted the issues mentioned in the description were prominent.

'Linda is 31 years old, single, outspoken, and very bright. She majored in philosophy. As a student, she was deeply concerned with issues of discrimination and social justice and participated in anti-nuclear demonstrations.' Which is more likely?

Linda is a bank teller.

Linda is a bank teller and is active in the feminist movement.

Given the issues of the day, Linda's background fits plausibly in Option 2 and that was the most frequent choice – even though it is obviously wrong. The joint probability (bank teller *and* active …) of two events must be less than the probability of either individual event. However, not even statisticians have immunity from System 1 thinking. If we slow down and use System 2, we are more likely than the general population to get this question correct, but our natural tendency is no better than the general population. The notions of coherence, plausibility, and probability are easily confused by System 1 thinking,

5.2.2.8 Endowment Effect

The endowment effect causes us to value an object we own or use more than an object we do not own or use. Owners assess the value of their house as higher than other similar houses. In business, the tendency is to value our product or portfolio of products higher than others; therefore, external projects or opportunities are undervalued compared to internal projects; or, projects in our part of the business are valued more highly than projects in another part of the business within our company.

5.2.2.9 Framing

How a problem is presented (framed) influences our choices. An especially important aspect of framing is establishing a default position. More employees invest in 401K retirement plans when the default is to contribute and they must opt out if they do not wish to contribute. Cafeteria customers make healthier food choices if salads are at eye level early in the traffic flow while candy bars and cookies are set off to the side. In these examples, each person is free to choose whatever they want, but decisions can be skewed toward 'smarter' choices.

In health care, doctors prefer an intervention that leads to a one-year survival rate of 90% versus one that has a one-year death rate of 10% – even though the outcomes are identical. In business, the decision of whether a project should be advanced to the next stage or phase of development is often unconsciously framed. The decision to advance a project is typically referred to as the 'Go, No go' decision, and development teams are often instructed to 'plan for success.' This framing, essentially another aspect of anchoring, sways deciders to the 'Go' decision and may lead to overestimates of the probability of success, and therefore higher than expected failure rates in the next phase of development.

5.2.2.10 Halo Effect
The halo effect is the tendency to like or dislike everything about a person, place, or thing, including what has not been observed. For example, funny people are often considered smarter than would otherwise be the case. A good first impression positively colors later negative impressions. In drug development, strong efficacy data tends to mask safety considerations for sponsors and safety concerns tend to mask strong efficacy data for regulators.

5.2.2.11 Hindsight Fallacy
The hindsight fallacy causes us to think we understand the past more so than is the case, which in turn causes us to think the future is more certain than it is. After an event occurs, we forget what we believed prior to that event. For example, 'I knew that product was going to be a dud in the market.' The result is that the hindsight fallacy fuels overconfidence in our ability to predict success or failure of the next project.

5.2.2.12 Loss Aversion
We will work harder and/or accept a greater risk to avoid losing $1000 than to win $1000. This makes it harder to terminate a project and makes it easier to give the project one more chance.

5.2.2.13 Motivational Bias

Why does the home crowd at a sporting event complain when a referee's call goes against the home team and almost never complains when the call goes against the visiting team? Because we root for the team we want to win, and this clouds our judgment. In business or research, we want to keep our project, compound, or product line alive, and this motivation can bias us to value our project over others that are in truth 'better.'

5.2.2.14 Overconfidence Bias

Psychologists have shown that humans are systematically overconfident. For instance, research shows that physicians are overly optimistic in their diagnoses, even if they have a great deal of experience. An example of drug development is the tendency to overestimate the probability of success of a drug, development plan, or study.

5.2.2.15 Priming

Things outside our awareness influence thinking. Conscious or subconscious exposure to an idea primes us to think about an associated idea. For example, if asked to fill in the blank SO_P, answers can be swayed toward U for SO<u>U</u>P if just prior to asking we have been discussing food, but if we have been talking about hygiene answers are swayed to A for SO<u>A</u>P. If you were asked to guess my age, answers could be swayed younger if just prior we had discussed children or swayed older if we had discussed grandparents.

5.2.2.16 Prospect Theory

This is Kahneman and Tversky's most famous work (Kahneman, 2011). One aspect of prospect theory is that framing matters. Even small changes in wording can have a substantial effect on the propensity to take risks. Their work shows that we make different decisions given alternative frames, even if the expected values in both situations are identical. Framing situations in terms of preventing loss (preventing deaths) promotes decisions with greater risk/uncertainty, whereas framing situations in terms of gains (saving lives) promotes decisions with less risk/uncertainty.

This idea has been extended to business decisions framed as opportunities versus threats. According to prospect theory, organizations act rigidly when faced with threats and act more flexibly and adaptively if the same situation is framed as an opportunity. For example, the tendency is to 'try harder' using well-established routines and procedures for a threat. This may, however, only continue the same things that led to trouble in the first place.

5.2.2.17 Recency Effect

The recency effect is a specific form of availability bias. The availability bias is when too much emphasis is placed on the information and evidence that is most readily available. The recency effect is when this availability bias stems from too much emphasis on recent events. In one study of decision-making in chemical engineers, scholars showed how the engineers misdiagnosed product failures because they tended to focus too much on causes that they had experienced recently.

5.2.2.18 Sunk-Cost Bias

The sunk-cost bias is the tendency to escalate commitment to a course of action in which substantial prior investments of time, money, or other resources have been made. If we behaved rationally, we would make choices based on the marginal costs and benefits of our actions. We would ignore sunk costs. With high sunk costs, however, we tend to continue the activity even if the results are poor. We 'throw good money after bad,' and the situation continues to escalate. Many business examples exist where a product was continued to be developed when the data suggested it should be terminated because 'We have invested so much already, we can't turn back now, we have to give it one more try.'

5.2.2.19 Theory-Induced Blindness

Once a theory has been adopted and used as a tool to guide thinking, it is hard to notice and accept its flaws. Common examples of how theory-induced blindness can make it difficult to rethink an idea include that the world is flat and that the earth is the center of the universe. In science we propose, entertain, and/or accept a theory, for example on how a drug can treat a disease, before we begin development, which can make it difficult to accept unfavorable results later in development.

5.2.3 Intuition

Some researchers and business executives are said to have great instincts, or a gut feel, for the areas in which they work. Somehow these experts just seem to know which products will succeed and which ones will not. As a youngster, I was a baseball fan and marveled at the centerfielder for my favorite team, Curt Flood of the St. Louis Cardinals. Flood was known for his ability to 'get a jump on the ball,' seeming to start running in the direction the ball was hit – before it was hit! Chess players 'see the board,' meaning they can anticipate moves far in advance. In *Thinking Fast and Slow*, the story is told of a Fire Chief who instructed his crew to evacuate a building in which there was a minor fire moment before the floor collapsed into the unrecognized inferno below, saving the lives of many in his crew.

How can such instincts be developed? The paragraphs below discuss what instinct is, and how and when it can be developed. Spoiler alert: Instinct can be useful in areas where we get frequent practice with rapid feedback, but these are not the conditions seen in many areas of science and business.

Intuition is not a magical ability or sixth sense. It is knowing something without knowing how we know it. Instinct is pattern recognition – being so familiar with something we arrive at accurate judgments under uncertainty using Kahneman's metaphoric System 1 (Kahneman, 2011).

Kahneman (2011) summarizes that instinct and intuition can be trusted when the environment is regular and predictable, and these regularities have been learned through prolonged practice with quick feedback. In the case of Curt Flood, by the time he was a Gold Glove award-winning center fielder for the Cardinals, he had played thousands of games, with batters making contact dozens of times with over 100 pitches per game. Feedback on where the ball was going to be hit was instantaneous. Chess experts have played thousands of games, with dozens of moves per game. They know within seconds to minutes whether their decision, their move, was successful or not.

Kahneman (2011) notes that although the Fire Chief who saved his crew by evacuating the building in the nick of time first credited his decision to instinct, upon further questioning the Fire Chief came to realize that his decision was based on pattern recognition. The key clues were that although the fire appeared to be minor, the floor was hot. This pattern did not fit a minor fire. The Chief did not know there was an inferno below, but he knew something was not right. He did not fully understand what was going on, and the experience of fighting fires of various origins nearly every day for decades had taught him that when the characteristics of a fire were not understood the best action was to get out of harm's way and further assess the situation.

In the clinical phases of drug development, even the most experienced researchers at the biggest companies may make decisions on only a few dozen drugs over their career, and it takes years to get feedback, to know whether the drug was successful. And when a drug is terminated, we almost never learn whether that was a false negative decision. Similar conditions exist in many areas of science and business. Infrequent decisions and slow feedback are not conditions where instinct can be trusted. Therefore, if someone has the reputation for having a great gut feel for business decisions, a more likely explanation is that they have been lucky.

Whether basing decisions on intuition or analysis, we often rely on reasoning by analogy. Analogical reasoning is assessing a situation and then comparing it to a similar situation. We consider what worked and did not work in the past. Based on that assessment, we decide what to do in the current situation (Kahneman, 2011).

Analogical reasoning can be powerful. It saves time because we do not start from scratch when searching for solutions to complex problems. Analogical reasoning allows us to use experience and avoid repeating mistakes, or to reuse what has worked well. Research also shows that some of the most innovative ideas come when we think outside of our field of expertise and make analogies to situations in different domains (Kahneman, 2011).

However, as is made apparent from understanding cognitive biases, representativeness and analogical reasoning can lead us astray. Research shows that we tend to focus on the similarities between the two analogous situations and downplay or ignore the differences. We can become enamored with analogies because aspects that do not fit the current situation are overlooked. The assumptions embedded in analogical reasoning are often not clear and therefore difficult to validate. For example, the factors that must line up for a past situation to inform the current situation may not be clear. Therefore, important differences may go unrecognized, and the past situation may be misleading. Therefore, when basing a current decision on a related situation, it can be useful to make detailed lists that describe how the scenarios are alike and different (Roberto, 2013).

5.2.4 Quantitative Analysis and Probability

Statisticians are well positioned to utilize quantitative analysis and probability in critical thinking and make decisions under uncertainty. However, that is not a guarantee of good decision-making. Consider the well-known Monty Hall problem described below which is based on the American television game show *Let's Make a Deal* and named after its original host, Monty Hall.

Suppose you are on a game show, and you are given the choice of three doors: Behind one door is a car; behind the other two doors are goats. You pick a door, say No. 1. Next, the host, who knows what is behind each door, opens another door, say No. 3, which has a goat. She/he then asks, 'Do you want to keep your original choice or switch to door No. 2?' Is it to your advantage to switch doors? The answer is provided below, but please write down your answer before reading further.

Switching yields a 2/3 probability of winning the car, while retaining the original choice has a 1/3 probability of winning the car. Most people think the probability of a car being behind the two remaining doors is 50%, and therefore conclude that switching does not matter. In one study with 228 subjects, only 13% chose the correct answer that switching was the best choice. When the Monty Hall problem and its solution were presented in the *Parade* magazine, many readers refused to believe switching is beneficial. Approximately 10,000 readers, including nearly 1,000 with PhDs wrote to the magazine, most of them insisting the stated solution was wrong. Even when given explanations, simulations, and formal mathematical proofs, many people still did not accept that switching is the best strategy (Roberto, 2013).

The basic explanation is that when the player first makes their choice, there is a 1/3 chance that the car is behind the chosen door and a 2/3 chance the car is behind an unchosen door. These probabilities do not change after the host reveals a goat behind one of the unchosen doors. There is still a 1/3 chance the car is behind the chosen door and a 2/3 chance the car is behind an unchosen door – but now there is only one unchosen door; hence, by switching to the unchosen door the probability of winning increases from 1/3 to 2/3.

Given how frequently educated and mathematically savvy individuals make the wrong decision, the problem has attracted the attention of cognitive psychologists. The reason even the smartest among us are frequently wrong is tied to the cognitive biases discussed earlier in this chapter. We use System 1 thinking and are therefore prone to the endowment effect, which causes us to overvalue the door initially chosen; the commitment or status quo bias, which causes us to stick with a choice we have already made; and the errors of omission versus errors of commission effect, in which, all other things equal, people prefer to make errors through inaction (retain choice) as opposed to making errors through action (switching).

5.3 PUTTING THE PRINCIPLES INTO PRACTICE

The key to practical applications of critical thinking and decisions under uncertainty begins with maintaining the orientation of individual, group, and organizational level factors. Group and organizational factors are covered in Chapter 6. Regarding individual-level factors, remember that humans – all of us, not just others – are not wired for making good decisions under uncertainty. Cognitive biases are the primary source of poor decisions and even the most intelligent scientists are prone to these biases. Of course, becoming more aware of cognitive biases and how they influence thinking can help. However, being aware of bias does not correct it. Specific plans, i.e., processes, are needed to prevent biases from creeping into our thinking. Table 5.1 provides a checklist of items to help prevent cognitive biases.

In decision-making, it is important to distinguish between confidence and accuracy. Confidence can stem from a story that

TABLE 5.1   Tactics to Help Prevent Cognitive Biases

Use multiple frames or anchors to guard against framing and anchoring bias
Ensure all relevant data are included to avoid the confirmation bias
Solicit external input to avoid endowment and motivational biases
Focus on marginal/incremental cost to avoid the sunk cost bias
Do not overlook objective data or statistical information in favor of gut feel because intuition is often wrong. Utilize algorithms whenever available rather than intuition

comes easily to mind, with no contradiction and no competing scenario. However, ease of recall and coherence do not guarantee that a belief held with confidence is true. In fact, research has shown that greater confidence is associated with poorer decisions because confident people have tended to overlook crucial aspects of the situation.

Conclusion drawn from intuition (System 1) fuels overconfidence. Just because something 'feels right' (intuitive) does not make it right. System 2 is needed to slow down and examine our intuition, estimate baselines, consider regression to the mean, and evaluate the quality of evidence. Therefore, confidence is a poor measure of accuracy and is not a reasoned evaluation of the probability of being correct. Instead, confidence is a feeling that reflects the coherence of the information and the cognitive ease of processing. It is therefore a mistake to base the validity of a judgment on the subjective experience of confidence rather than on objective facts and statistical evaluation.

5.4 REAL-WORLD EXAMPLES

The following problem is common in the pharmaceutical industry, with similar applications in many business and scientific settings. This example illustrates the value of rigorous quantitative analysis in making decisions under uncertainty.

If a proof-of-concept clinical trial is conducted using a design and analysis that controls the probability of a false positive result and a false negative result at 10% for each, and the study result meets the critical success factor of $p < 0.10$, all else equal, what is the probability the drug is effective?

Based on classical frequentist hypothesis testing alone, the probability of a false positive result from the trial is 10%. Therefore, many will interpret this result as suggesting a 90% probability that the drug is effective. Of course, statisticians, especially those with a Bayesian perspective, recognize that the result from frequentist hypothesis testing may be misleading because the background rate should also be considered. Using a background rate typical of drug development, 10% of drugs that enter proof-of-concept testing are in truth effective, the probability the drug in this hypothetical scenario is effective is approximately 50%, not 90%.

However, as discussed in Chapter 4, we should explain the reasoning behind this result to non-statisticians because they lack the mathematical knowledge to understand the Bayes theorem. The following example could be a useful, non-statistical explanation. Say we have 1,000 drugs in a hypothetical portfolio to be tested in 1,000 proof-of-concept studies, with each study having a 10% probability of a false negative result (90% power) and a 10% probability of a false positive result (based on frequentists' hypothesis testing). Therefore, the hypothetical portfolio has 100 drugs that are in truth effective and 900 that are ineffective.

100 effective drugs tested at 90% power yields 90 true positives and 10 false negatives, and 900 ineffective drugs tested with a 10% false positive rate yields 90 false positives and 810 true negatives.

Therefore, the results include 180 positives and 820 negatives. Of the 180 positives, 90 are true positives, yielding a 50% probability that a drug with a positive proof-of-concept test is in fact effective.

Although the arithmetic involves several steps, it is more likely non-statisticians will understand and accept this result than from an explanation based on the Bayes theorem. Hence, this is a useful illustration of how statisticians and the quantitative rigor we bring can be an essential component to critical thinking and making decisions under uncertainty.

Nevertheless, the Monty Hall problem covered earlier in this chapter is a humbling reminder that even mathematically savvy scientists can struggle with critical thinking and decisions under uncertainty.

II

Working with Others

CHAPTER 6

Critical Thinking and Decision-Making

Group and Organizational Factors

ABSTRACT

This chapter begins Part 2, Working with Others, and covers group and organizational factors that influence critical thinking and making decisions under uncertainty. Unfortunately, groups do not always harness the collective intellect and creativity of the individual members, and there is much more than group think that can get in the way. This chapter covers the characteristics of dysfunctional teams that can hinder critical thinking and decision-making, and discusses what can be done to foster highly functional groups that do harness the collective abilities of its members. This chapter also covers the characteristics of entire organizations that provide the environment which can also influence group dynamics. The chapter includes practical

DOI: 10.1201/9781003334286-8

tips on how to help groups function more effectively to enhance critical thinking and decision-making, and includes real-world examples from the author's career.

6.1 INTRODUCTION

Conventional wisdom is that groups can make better decisions than individuals because groups can pool the diverse talents of the team. Merging ideas from diverse perspectives can create the potential for new ideas and options that individuals could not create on their own. However, groups often do not realize this synergy because coaxing and merging ideas from multiple perspectives is often a difficult and lengthy process. These 'process losses' can be so substantial that the group does worse than the best individual in the group could do on his or her own (Roberto, 2013).

Group think is a well-known cause of flawed decisions. Group think can happen even if the group is cohesive, with smart individuals who all have good intentions. Roberto (2013) quotes social psychologist Irving Janis in stating that group think is more likely when a team is cohesive and experiences pressure for conformity and consensus, which comes at the expense of critical thinking. Roberto (2013) further cites Janis in noting that results of group think include discussions that lack diversity and divergent thinking. Risks associated with a plan are not surfaced because assumptions are not transparent and debated. Once an option is dismissed, it is rarely reconsidered to see if it could be bolstered and made more plausible. The group does not seek independent, outside experts. Confirmation bias may result from limited information. The group does not discuss contingency plans. Roberto concludes with a final quote from Janis that group think can be avoided by first 'deciding how to decide.'

Some key dimensions of deciding how to decide are (Roberto, 2013):

- Composition: Who should be involved in the decision-making process

- Context: The environment in which the discussion and decision takes place

- Communication: What are the 'means of dialogue' among the participants

- Control: Who will guide or lead the discussion

Team leaders are particularly important in deciding how to decide. Leaders should consider how directive they want to be regarding the content of the decision, the process of decision-making, and how much they want to control the outcome of the decision. If leaders retain too much control over the process, the end decision may reflect the thinking of the leader and not the collective wisdom of the group (Roberto, 2013).

Team members do not want a 'charade of consultation' where alternatives are developed and discussed, but the decisions were already made; hence, leaders steer the decision toward their preferred choice and then announce the decision as if it were the result of the current discussion (Roberto, 2013).

In a fair process, people:

- Have ample opportunity to express their views and to discuss how and why they disagree with others

- Feel that the decision-making process was transparent (i.e., free of behind-the-scenes maneuvering)

- Believe that they have been heard and their views considered thoughtfully before the decision is made

- Have a genuine opportunity to influence the final decision

- Have a clear understanding of the rationale for the final decision

However, these goals are not easy to attain (Roberto, 2013). *The Five Dysfunctions of Teams* (Lencioni, 2002) presents a pyramid

TABLE 6.1   The Five Dysfunctions of Teams

Lack of trust
Fear of conflict
Lack of commitment
Lack of accountability
Lack of focus on results

model for team dysfunction in which one dysfunction leads to another, leading to low morale and poor decisions. The five dysfunctions are listed in Table 6.1.

At the bottom of the pyramid is the absence of trust. With this dysfunction, team members are reluctant to show their weaknesses and vulnerabilities, and are not open and honest with one another. This results in team members being afraid to admit mistakes and an unwillingness to ask for help. Lack of trust leads to fear of conflict, which in turn results in team members being incapable of engaging in debate or voicing their opinions (Lencioni, 2002).

Fear of conflict leads to a lack of commitment because team members do not buy into the inferior decisions that result from the lack of robust discussion. Because team members lack commitment, they do not hold each other accountable. If team members do not feel accountable, they put their own needs (e.g., ego, recognition, career advancement) ahead of the team goals. This results in the team losing sight of goals, and the results suffer. Using this pyramid model helps make clear that group think is at least as much a symptom of team dysfunction as it is a root cause (Lencioni, 2002).

As the first step for addressing these dysfunctions, Lencioni notes that for teams to understand that these dysfunctions exist in most situations where the team does not actively work to counteract their natural tendencies. Team leaders should lead by example and be the first ones to admit weaknesses and vulnerabilities, to encourage debate and challenges to their thinking, making responsibilities and deadlines clear, setting team standards, and being clear on the team's outcome. Leaders should

TABLE 6.2   Diagnostic Questions to Assess Level of Team Dysfunction

Do team members openly and readily disclose their opinions?

Are team meetings compelling and productive?

Does the team come to decisions quickly and avoid getting bogged down by consensus?

Do team members confront one another about their shortcomings?

Do team members sacrifice their own interests for the good of the team?

not, at least initially, focus on reaching a consensus. Instead, the initial focus should be to make sure that everyone is heard (Lencioni, 2002).

Lencioni (2002) advocates asking the questions in Table 6.2 to begin understanding the level of dysfunction on a team:

No team is or can become perfect. However, working to ensure that the answers to the above questions are yes will go a long way to making a team as good as it can be in critical thinking and making tough decisions in uncertain situations.

An important aspect embedded within Lencioni's ideas on team function is the benefit of evaluating multiple alternatives. A decision can be no better than the best alternative. In drug development, a good example is deciding on a study design or a clinical development plan. All too often, the tendency is to decide on the design early in the process so that the hard work to follow can begin. Similar ideas apply to many business applications (Lencioni, 2002).

An explicit idea-generation phase of discussions can lead to alternatives that may otherwise have not been identified. For example, a good way to begin discussion on how to design a study or product XYZ is not to ask team members how they think the study or product should be designed. Instead, begin the discussion by asking the team to develop three options. Option 1 may be the fastest plausible option; Option 2 the cheapest plausible option; and Option 3 is the option with the greatest potential value or chance of success.

By starting the debate in this way, no one puts a stake in the sand – takes a personal position – from which it would be hard to convince them to change their mind later. In fact, it is best to suppress personal preferences at this stage. In doing so, everyone feels invested in the alternatives. The next step may be to list the strengths and limitations of each plan, many of which will be self-evident, but some may be less clear. With this background in place, then ask the team to come up with other options that capitalize on the strengths and limits or mitigate the weaknesses of the previously identified plans. With this approach, the entire team is invested in the process and therefore more likely to be committed to the final decision.

6.2 ORGANIZATIONAL FACTORS THAT INFLUENCE CRITICAL THINKING AND DECISION-MAKING

The Art of Making Critical Decisions (Roberto, 2013) notes three types of organization cultures that can be problematic in decision-making. Although most statisticians will not have roles in which they can influence the culture of an organization, it is nevertheless useful to understand these cultures. The three dysfunctional cultures are the culture of no, the culture of yes, and the culture of maybe.

The 'culture of no' is a phrase coined by Lou Gerstner, CEO of IBM. Sometimes, a culture of no arises in an organization because meetings have become places where participants strive to deliver 'gotchas,' or 'smart talk.' That is, some organizations reward those who are great at dissecting others' ideas, even if they offer no alternatives themselves. Individuals are rewarded for 'gotcha' moments, as if they were still in an MBA classroom.

In the culture of yes, people tend to stay silent if they disagree with a proposal. Then, after the meeting they express disagreement, lobby to overturn the choice, or undermine the implementation of the plans with which they disagree. This leads to the appearance of consensus at meetings when consensus does not exist. A key takeaway is that silence does not mean assent. When people are not contributing to a discussion, they may disagree but not wish to voice their dissent in the meeting.

A culture of maybe entails the desire to gather as much information as possible, so much so that they get caught in 'analysis paralysis.' Decisions or actions are delayed because a bit more information and analysis might be beneficial. The culture of maybe afflicts people and groups who struggle to deal with ambiguity, or who engage in conflict avoidance. Indecisive cultures often originate from past success wherein some variant of dysfunctional behavior worked in the past, or at least success was achieved despite the dysfunction. Hence, the behavior is repeated.

Roberto (2013) further notes that most of the failures in complex organizational decision-making do not have a single cause. They typically involve a chain of decision failures, a series of small errors that are built upon one another in a cascading effect. One decision failure that would be insignificant on its own leads to additional errors that cumulatively lead to failure.

Roberto (2013) describes the Swiss cheese analogy as thinking about how organizations can limit the risk of these decision failures. In this depiction, an organization's layers of defense or protection against accidents is like slices of Swiss cheese, with the holes in the block of cheese representing the weaknesses in defenses. In most instances, the holes in a block of Swiss cheese do not line up perfectly, such that one cannot look through a hole on one side and see through to the other side.

In other words, a small error may occur, but one of the layers of defense catches it before it cascades throughout the system. However, in some cases, the holes become aligned such that an error can slip through and cascade through the organization. Reason argues that we should try to find ways to reduce the holes (i.e., find the weaknesses in our organizational systems) as well as add layers (build more mechanisms for catching small errors).

Normalization of deviance is another source of organizational failure (Roberto, 2013). It is the phenomenon wherein over time more risk is taken on or greater deviation from the stated or understood plan occurs. Roberto (2013) uses the example of Enron to illustrate how normalization of deviance can take down an organization. Managers took on increasing risk as they pushed for the growth needed to meet the expectations of top management and Wall Street. Enron did not suddenly step over the line from normalcy to deviance. They normalized the risks over time. They moved ever further away from their core business in natural gas, taking on ever greater risk.

Roberto (2013) goes into more detail on additional aspects of organizational decision-making. Interested readers are encouraged to study these ideas. However, for the present purposes, we close with the key conceptual takeaway. We cannot simply look at the leader to understand decision-making in complex organizations. Decisions are not the product of the leader's cognitive process. We all play a part, however small. And even if we feel we cannot influence the entire organization, which is especially true at large companies, understanding these ideas can provide insight into how our company works, and thereby make us a more valuable contributor.

6.3 QUALITY DECISION-MAKING FRAMEWORKS

6.3.1 Introduction

Many of the ideas in the previous sections of this chapter can be put into action via what is referred to in the project management world as the Quality Decision-Making Framework, or QDM. Decision quality concepts were first developed in 1964, building on advancements in statistical decision theory and game theory by Howard Raiffa of Harvard University, and dynamic probabilistic systems by Ronald A. Howard of Stanford University. The first implementation of these concepts in business was documented in decision analysis: Applied decision theory (Howard, 1966). Since then, decision analysis tools and decision quality concepts have been adopted by many corporations to guide and improve their decisions.

Fundamental to all decision quality concepts is the distinction between the decision and its outcome (Spetzler et al., 2016). The distinction is important because high-quality decisions can

TABLE 6.3   Elements of the Quality Decision-Making Framework

Framing
Options
Information
Values and trade-offs
Sound reasoning
Commitment to action

still yield poor outcomes, and vice versa, due to uncertainties. In the face of uncertainty, the decision-maker has control over the decision, but does not have control over external circumstances, chance, and other potentially unknown factors. Consequently, the outcome of a decision does not allow an assessment of its quality. A decision has quality at the time it is made, which is not changed by hindsight. Concepts of decision quality focus on measuring and improving the quality of the decision at the time it is being made. The six tenets of the QDM process and quality decisions are listed in Table 6.3 and described in detail after the table (Howard, 1966).

6.3.1.1 Framing

The first element to achieve decision quality is framing. Having the appropriate frame ensures the right decision problem is addressed. Quality in framing is achieved when the decision-makers align on purpose, perspective, and scope of the problem to be solved. It means the right people will work on the right problem in the right way.

6.3.1.2 Options

A decision cannot be better than the best available alternative/option. A wide variety of approaches, tools, and methods exist to generate high-quality options. Quality in options is achieved by applying a suitable options generation process to produce a diverse set of feasible options, some of which may be hybrid solutions that combine the best features from multiple options.

Developing feasible options implies an understanding of implementation.

6.3.1.3 Information

The quality of a decision depends on the quality of the information to inform the decision. Quality in information is achieved when the information is meaningful and reliable; is based on appropriate data and judgment; reflects all uncertainties, biases, intangibles, and interdependencies, and the limits to the information are known. A wide variety of tools exist to improve the quality of the information used in the decision problem.

6.3.1.4 Values and Trade-Offs

Quality in this element requires identifying the right decision criteria and defining the trade-offs among them, which in turn necessitates first identifying the key stakeholders, and what each of them values. Quality in this element is characterized by transparent metrics, a clear line of sight of the primary metric, and explicit trade-off rules between key metrics.

6.3.1.5 Sound Reasoning

This element is the domain of decision analysis, which aims to produce insight. Decision analysis provides the logic and analytic tools to find the best choice in a complex situation and is a guide in facilitating conversations about the decision. A wide variety of tools, ranging from decision trees to complex network models, is available to match the decision problem. Quality in this element is achieved when the value and uncertainty of each alternative are understood, and the best choice is clear.

6.3.1.6 Commitment to Action

The quality of a decision also depends on the commitment to act upon the choice that is made. Quality in this element is achieved by involving all key decision-makers and stakeholders in an effective and efficient decision-making process. At the end of the process, quality is characterized by buy-in across all stakeholders and an organization that is ready to act and commit resources.

6.4 PUTTING THE PRINCIPLES INTO PRACTICE

Groups can avoid inadvertent framing bias by holding back on any individual stating their opinion on the decision until all relevant information has come to light. This is especially true for leaders or senior team members because if they express an opinion it tends to stifle further debate. It is also useful to adopt multiple frames when examining a situation. In other words, defining problems in several ways is useful because each definition (frame) tilts us toward one kind of solution. Explicitly specifying assumptions, including implicit assumptions, and testing those assumptions and their impact on inferences are key to making good decisions and avoiding framing bias.

Groups, like individuals, need to distinguish between confidence and accuracy. As described in Chapter 5, confidence can stem from a story that comes easily to mind, with no contradiction and no competing scenario. However, ease of recall and coherence do not guarantee that a belief held with confidence is true. Remember, research has shown that greater confidence is associated with poorer decisions because confident people have tended to overlook crucial aspects of the situation.

Conclusions drawn from intuition (System 1) fuel overconfidence. Just because something 'feels right' (intuitive) does not make it right. System 2 is needed to slow down and examine our intuition, estimate baselines, consider regression to the mean, and evaluate the quality of evidence. Hence, the group needs to move at a pace and in a manner that allows its members to think via System 2.

Some ways to avoid group think include role-playing and mental simulation methods to stimulate conflict and debate. An example in the business world is to use structured methods to envision and plan for multiple future scenarios in the marketplace. Debate can be stimulated with point-counterpoint methods and the use of devil's advocates. Although a designated devil's advocate can be useful, the status as a designated role can dilute its power because others know the advocate lacks conviction in the stated beliefs. Therefore, a true devil's advocate is especially useful.

Leaders and senior team members should be watchful for the signs of unproductive debate listed in Table 6.4.

It is important to understand the difference between the two forms of conflict. Cognitive conflict is task oriented. It is a debate about issues and ideas. Affective conflict is emotional and personal. It is about personality clashes, anger, and personal friction. Leaders and senior team members should try to stimulate cognitive conflict while minimizing affective conflict. This is a more explicit statement of the advice to play hard but play nice.

Constructive conflict can be stimulated by ground rules for how people interact during the deliberations. For example, it can be useful to clarify the role that individuals will play in the discussions. People can be assigned to play roles that they usually do not play to help them understand how their colleagues think about problems.

Leaders can divide the group into sub-teams to organize arguments for and against a proposal without regard for the individual opinion. Leaders can redirect attention and recast the

TABLE 6.4 Signs of Unproductive Team Debate

People have stopped asking questions to gain a better understanding of viewpoints
The group has stopped searching for new information
Individuals have stopped revising their proposals based on feedback and critiques
No one is asking for help with the interpretation of ambiguous data
Repeating the same arguments
No one admits concerns about their own proposals
Less outspoken individuals withdraw from the discussions

situation in a different light, which is to reframe the situation. After-action reviews can help improve a group's ability to manage conflict constructively (Roberto, 2013).

Creativity requires a willingness to focus on avoiding premature convergence of a single idea. Judgment should be deferred to generate diverse ideas. Leaders and senior team members should be willing to experiment and fail, to provide ideas, and to let them be shot down. The willingness to fail and encouraging others to make useful and intelligent mistakes is critical to getting to the best decision (Roberto, 2013).

Table 6.5 lists some warning signs leaders and senior team members should watch for to indicate that the group does not have candid dialogue in recurring meetings (Roberto, 2013).

Leaders and senior team members can create barriers to candid dialogue if (Roberto, 2013):

- Team members' roles are ambiguous

- The team is too like-minded

- Large status differences among team members exist

- Leaders and senior team members present themselves as infallible and fail to admit mistakes

TABLE 6.5 Warning Signs That Indicate Possible Lack of Candid Group Dialogue

Meetings are usually polite monologues rather than strong debates

Subordinates wait to take their cues from leaders and senior team members before commenting on controversial issues or difficult problems

Planning and strategy sessions focus on pre-work and refining presentations rather than focusing on open dialogue

The same people tend to dominate the meetings

It is rare for the leader to hear concerns or feedback directly from subordinates

Management meetings become 'rubber stamp' sessions in which leaders ratify decisions that have already been made through other channels

6.5 REAL-WORLD EXAMPLES

The examples presented in this section are not presented as firm recommendations for what to do in similar situations. They are examples of how groups I was involved in made decisions. These examples may or may not be useful in situations relevant to you.

6.5.1 An Example of Figuring Out What to Do

A team I was supporting was stuck and could not align on the best plan for Phase 3 development. The team had received unexpected feedback from Food and Drug Administration (FDA), which necessitated a major change in development, but the team could not settle on a plan to best implement the FDA feedback. Discussion proceeded through regular team meetings for six weeks with no substantive progress. Each weekly meeting covered the same points.

Because the team had gathered enough information but was having trouble processing that information, a new approach was needed. Therefore, a structured one-day 'Development Workshop' was scheduled. Two of the team's most experienced researchers put together a brief pre-read to summarize FDA feedback and other relevant information, along with a high-level description of two general paths forward (Plan 1 and Plan 2). Eight functional experts were divided into two sub-teams, which I will call Team A and Team B.

Each team was given the first two hours of the workshop to separately develop details for a plan. Team A worked on Plan 1 and Team B worked on Plan 2. Then teams rejoined with each sub-team taking 15 minutes to highlight their plan. Then Teams switched plans. Team A took Plan 2 that was developed by Team B with the task of making the plan better – not changing fundamental aspects of the plan or criticizing the plan – but tweaking it to make it better. Similarly, Team B took Plan 1 that was developed by Team A and tweaked it. After 90 minutes, the teams rejoined. The next hour was spent discussing the strengths and limitations of the refined plans. The Therapeutic Area Leader

– the key decider – was invited for this hour and for the remainder of the workshop. After discussing the strengths and limitations of the plans, the full team was expected to vote on which plan they favored.

However, the workshop facilitator had, unbeknownst to the teams, set aside additional time thinking that the discussion of strengths and limitations might lead the team to a third plan, an optimum blend of the two plans. That is exactly what happened.

After discussing the strengths and limitations of the two plans, a third, optimum plan became clear. The full team cast secret ballots to rank order their preference on plans. Plan 3, the new hybrid, was the number 1 choice on six of eight ballots, and the second choice on the other two. With this strong alignment, it was straightforward for the Therapeutic Area Head to bring forward to the company governance plan 3.

Some of the reasons this workshop was successful included that by switching plans back and forth between teams, everyone felt ownership of the end results. It was not a competition. The cooperative attitude was reinforced by having sub-teams build on each other's plans rather than critiquing them. Having time-bounded discussions was also useful. Although decisions cannot be forced, in this case, lots of discussion had already happened and the team benefitted from a deadline to hone thinking.

In considering the success of the workshop, the team had originally been trying to come up with a perfect plan in team meetings. The plans that seeded the workshop discussions were not perfect, but provided a useful starting point, from which the structured discussions could build.

6.5.2 An Example of Choosing between Competing Plans in a Group Setting

Early in my career, I was supporting a drug that was initially to be developed based on BID (twice daily) dosing to minimize adverse effects, with subsequent bridging to QD (once daily) dosing. However, revised input from marketing indicated that the drug needed to be launched with QD dosing. Two plans emerged, with a roughly equal split in support for the two plans among team members. The team could not come to a consensus throughout repeated deliberations. Each plan had compelling strengths and limitations, which were clear to all. Further debate was only strengthening each camp's support for their plan.

To break the deadlock, the team leader called a meeting at which each camp would have one last articulation of their position, with the strengths and limitations of each plan documented. The team leader stipulated that at the completion of the meeting all discussion would cease, and the next morning the team leader would render his decision, which would be final and fully supported by all.

The team leader did not choose the plan I favored; he chose the opposing plan. Despite the long period of disagreement, there was no animosity after the decision. We implemented the chosen plan effectively and the drug was successfully launched with QD dosing. This approach was effective because everyone felt they had been heard and it was clear the team leader based his decision on what was best for the team, not based on individual favoritism.

6.5.3 An Example of Two Individuals Coming to Alignment

An early-phase statistics group leader had a direct report who was the lead statistician for a compound that was planning a Phase 2 dose-ranging study. The direct report, I'll call her Claire, and the manager, I'll call him Fred, disagreed on how many doses should be tested in the Phase 2 study. Fred favored testing four active doses and placebo over a wider range, Claire favored testing three active doses and placebo over a somewhat narrower range. It was important that the statistics function present a united front to the rest of the team. Competing statistical viewpoints would be a distraction and the differences needed to be resolved in advance.

Fred called a meeting where he and Claire laid out their arguments. Claire went first and then Fred presented his case. After Fred was finished, Claire said, 'OK then, I guess we will do it your way.' Fred said,

> No, that is not what we are going to do. Although you have not convinced me that testing 3 doses is better than 4, the clarity and conviction of your arguments did convince me that you have thought about this carefully and that you know what you are talking about. You work with the team every day, and know the preclinical and Phase 1 data inside and out.

Fred explained to Claire if she had made some mistake or overlooked something important that he would have enforced his plan. However, because Claire had shown strong knowledge of all the important issues and the data, and had rigorously considered Fred's input, Fred supported her decision. The result was that Claire felt great ownership of the study and she continued to work with great rigor to make the study a success.

CHAPTER 7

Influence and Leadership

ABSTRACT

This chapter covers influence and leadership. Being an effective leader requires the ability to influence others. However, the chapter shows how influence is important for all levels of statisticians, not just statisticians in leadership, managerial, or senior technical positions. The chapter begins by dispelling five myths about leadership and then covers the basic principles of influence and power. A key focus of the chapter is on what influence tactics can be used with the various sources of power, and how likely those tactics are to be successful. This chapter then discusses transactional and transformational leadership, along with the fundamental principles associated with each. The chapter makes the important point that while some of us have a greater inherent ability to influence and lead, we can all improve and no one is so good that they are born with all the skills. These points are reinforced via examples from the leadership experiences of several US presidents. The chapter concludes with some practical advice

DOI: 10.1201/9781003334286-9

for putting influence and leadership principles into practice, along with some real-world examples from the author's career.

7.1 INTRODUCTION

Being an effective leader requires the ability to influence others. However, influence is important for all levels of statisticians, not just statisticians in leadership, managerial, or senior technical positions. Therefore, this chapter begins with basic principles of influence, including the psychology of influence, and a conceptual model for applying those basic principles. These basic ideas on influence provide the necessary background for the section that follows on key ideas on leadership.

7.2 INFLUENCE

7.2.1 The Psychology of Influence

In his classic book, *Influence, The Psychology of Persuasion* (Cialdini, 2008), Robert Cialdini explores the psychology of influence – why people say yes. Although this material is often presented in and applied to marketing contexts, the basic principles also apply to scientific contexts where we attempt to persuade others to believe in an idea or to take a specific action. Therefore, statisticians fluent in these principles are likely to be more effective in influencing their team, organization, and the scientific community.

Cialdini (2008) outlines the six universal principles of persuasion as listed in Table 7.1, each of which is explained in the sections that follow.

7.2.1.1 Reciprocation

Reciprocation is the desire or tendency to try and repay what another person has provided. Cialdini (2008) includes the following examples of reciprocation:

- If someone buys you lunch, you feel obligated to buy them lunch next time.

TABLE 7.1   The Six Principles of Influence

Reciprocation
Commitment and consistency
Social proof
Authority
Liking
Scarcity

- At the supermarket or warehouse club, 'free' samples encourage the reciprocity rule to get you to buy something you would not have otherwise bought.

- If a date takes you out to an expensive dinner, you feel obligated to go out with her/him again even though you were not that into them.

Scientists may consider themselves immune to these influences. However, research has shown no group is immune. Therefore, in scientific settings if a colleague has collaborated with us or endorsed our work, we are more likely to collaborate with them and/or endorse their work. To exploit this principle in scientific settings, a good way to gain support for an idea is to have earlier supported someone else's idea. This is not to say the support should be disingenuous. However, when consistent with our scientific understanding, showing prior support will make us more effective influencers.

7.2.1.2 Commitment and Consistency

This principle is about the desire to be (and to appear to be) consistent with what we have already done. Once a choice is made or a stand is taken, we encounter personal and interpersonal pressures to behave consistently with that commitment. Those pressures cause us to respond in ways that justify an earlier decision.

Cialdini (2008) uses the following examples to illustrate the principle of commitment and consistency:

- Staying married even though divorce may be the best option because we made a public commitment 'til death do us part'.

- One who has publicly stated her/his belief or theory is likely to continue to bring up the issue even if there is conclusive evidence countering the belief or theory

- If a runner announces they are running their first marathon in three months, they will be more motivated to put in the required training. The public announcement, or what Cialdini calls 'forced accountability,' is why stating or writing down goals increases the likelihood of achieving that goal

In scientific settings, commitment and consistency can be a two-edged sword. As with the marathoner, the principle of commitment and consistency can provide motivation to work on a long and difficult project. However, it can also make it more difficult to change our mind or to change the mind of others despite new evidence. Therefore, to capitalize on the commitment and consistency principle when presenting a new idea, we can point out how the new idea is consistent with someone's prior stance. Alternatively, we can point out how the current situation is different from previous situations, thereby mitigating the commitment and consistency principle that would otherwise cause someone to maintain an erroneous belief or continue a suboptimal course of action.

7.2.1.3 Social Proof
Social proof is often referred to as peer pressure or herd mentality. This principle influences what is considered correct behavior. If most people do it, it must be Ok. Cialdini (2008) uses the following examples to illustrate social proof:

- At a bar your friends order margaritas, so you do the same

- You start wearing attire similar to others

- You laugh at a joke because your friends are laughing, but you do not get it

- You see everyone staring up at the sky, so you look up too

Scientists often consider themselves free thinkers, immune to herd mentality. However, important examples to the contrary exist. For example, the last observation that carried forward the approach to imputing missing data in longitudinal clinical trials remained in widespread use long after its biases were well known. More generally, research suggests we consider others prone to herd thinking while erroneously considering ourselves immune. To exploit the principle of social proof, it can be useful to point out how our idea, recommendation, or course of action is consistent with other well-established ideas or actions.

7.2.1.4 Liking

Liking is a simple concept. We prefer to say yes to the requests of people we know and like. Therefore, it is important to understand what factors cause one person to like another person. Cialdini (2008) offers the following examples:

- At the time of this writing, Kim Kardashian has approximately 10 million 'likes' on Facebook. No doubt, having a reality show and a clothing line contributes to people knowing of her, but why so many likes? Evidence of unusual intelligence or talent is not obvious. However, she is widely considered attractive and fashionable; therefore, people want to be like her and in some way associated with her.

- We like people who are similar to us, whether it is sharing opinions, personality traits, background, lifestyle, etc.

- We enjoy and respond favorably to compliments.

- We tend to like things that are familiar and dislike what is unfamiliar.

- We like people who work with us, instead of against us. Working together towards a common goal and being 'on the same side' are powerful motivators.

- The principle of Association can drive either negative or positive connections as an otherwise innocent association with either bad things or good things influencing people's feelings. Everyone wants to be part of a winning team because it raises social standing. People will therefore try to link themselves to positive events and distance themselves from negative events. This is the root of sayings such as she/he was guilty by association, or she/he hung out with the wrong crowd. The principle of association is also the motivation behind name dropping.

Again, scientists may feel immune to these forces, but we are not. Therefore, attempts at influence will be more successful if we use genuine compliments, dress more consistent with current trends, and make our favorable associations known. Although on this last point, care is needed to not brag or condescend. However, it will not hurt to highlight affiliation with a revered institution on a CV or on the title slide of a presentation. Similarly, thanking an esteemed collaborator on an acknowledgement slide will give credit where credit is due and enhance your credibility.

7.2.1.5 Authority

People tend to follow authority figures. We are taught from a young age that obedience to authority is right, and disobedience is wrong. Cialdini (2008) includes the following examples:

- Police officers, firefighters, clergy, office managers, etc.

- Titles (PhD, Esq, MBA, etc.)

- In advertising, when the pitchman or pitchwoman is wearing the white lab coat of a doctor when selling toothpaste. Even someone acting as a doctor projects authority

Academic settings have obvious examples of authority in action. That is why academic credentials and awards are highlighted on CVs and in introductions to presentations.

7.2.1.6 Scarcity

The scarcity principle states that opportunities seem more valuable when their availability is limited. An example of scarcity is the loss aversion bias mentioned in Chapter 5, where the fear of loss is greater than the desire for gain of an equally valuable asset or amount of money.

Cialdini (2008) notes the following examples of scarcity that are frequently used in advertising:

- Limited time offers – A certain product is in short supply that cannot be guaranteed to last long: 'Order while supplies last' or 'order before prices go up.'

- Deadlines are another advertising attempt to evoke scarcity. Black Friday and Cyber Monday sales are examples.

The principle of scarcity works because things that are difficult to possess are typically better or more valuable than things that are easy to possess. An item's availability is often used as a quick proxy to evaluate its quality or value, rather than a more detailed evaluation of its merits or value. This is a classic example of using System 1 thinking rather than System 2 thinking (see Chapter 5).

Although connections between the principles of scarcity to endeavors of scientific influence are not obvious, it is useful to keep in mind the general idea of how easily our thoughts can be influenced independent of objective evidence.

7.2.2 Conceptual Model of Influence

In Kenneth G. Brown's book *Influence* (Brown, 2013), he outlines a conceptual model for influence that builds upon Cialdini's six principles of influence. Brown refers to this as the ATTC model

Outcome = $\underline{A}$gent + $\underline{T}$arget + $\underline{T}$actics + $\underline{C}$ontext

FIGURE 7.1   The ATTC model of influence.

(the acronym is ATTC, pronounced attic, as in the part of a house between the upper floor and roof). The model is illustrated in Figure 7.1.

In the ATTC model, the outcome of an influence attempt is a function of the $\underline{A}$gent, the $\underline{T}$arget, the $\underline{T}$actics, and the $\underline{C}$ontext. Outcomes in the ATTC model have three levels:

- Commitment: Willing and enthusiastic, needed for complex/difficult tasks

- Compliance: Willing but apathetic, minimal effort, works for routine tasks

- Resistance: Opposed to the request, actively tries to avoid doing it

In the ATTC model, the Agent is the person or entity doing the influencing, such as an advertising campaign, the marketing department of a pharmaceutical company, or scientists trying to convince colleagues of a new idea, theory, or result. The Target is the person or entity who is being influenced, such as consumers of products targeted by advertisers, or the audience at a scientific conference. Tactics are what the Agent does to influence the Target. Common tactics in the ATTC model are often based on the six principles of influence discussed in the previous section. Context is the situation and conditions under which the Agent and Target interact. Context can include consumers watching advertisements on TV, a pharmaceutical sales representative interacting 1:1 in a doctor's office, a scientific debate at a conference, or a journal article read by other scientists.

Because science should be based on objective evidence rather than emotion and bias, it is worthwhile to consider characteristics of bad agents so that they can be avoided at all costs in our scientific work. Characteristics of bad agents include taking advantage of others in business, life, and emotion (Brown, 2013).

Brown (2013) explains the dark triad – three characteristics typical of bad agents. The first characteristic is that bad agents tend to be Machiavellian; they are manipulative, sensitive to punishment, and narcissistic: Bad agents frame situations to be about them – the I complex. When bad agents fail, they do not own up to these mistakes, but instead make excuses for the failures, often blaming others. Bad actors also tend to have little sympathy for others, a trait referred to as psychopathy. Lastly, it is not uncommon for bad actors to utilize misinformation and to have malice toward others. The Target and Context of a situation can lead to greater potential for bad actors. Dictators and political tyrants are examples of bad agents in action. Being aware of the attributes of bad actors can help us to focus on being good actors, thereby using influence for good rather than personal gain (Brown, 2013).

Being aware of the characteristics of good agents can also enhance persuasiveness for positive purposes. Good actors have socially beneficial motives and use ethical approaches with accurate and complete information (Brown, 2013).

Brown (2013) describes the following characteristics of effective Agents, whether good or bad: Perception, appearance, charisma, apparent trustworthy behavior, competence, consistency, caring, and connecting with the Target as being on the same team. These attributes can be enhanced by using metaphors, stories, exuding confidence and passion, posture, gesture, and tone of voice. Being effective in these areas requires confidence built through preparation and practice. Most notably, acting positive builds positivity, which is charismatic.

Browne (2013) explains that Tactics include, but are more than, the six principles of influence. He lists the following examples of Tactics:

- The hard sell where the Agent pushes products or ideas on the Target

- Demands or threats, such as 'I'm the boss, do what I say or else.' As unappealing as this tactic seems in scientific settings, it is effective in crises, emergency room settings, and the military when there is no time for discussion

- Soft sells where the Agent shows the way or exemplifies the desired action

- Appeals to logic or emotion

Good agents listen to the Target and adjust the Tactics through understanding the Targets' perspectives. As such, flexibility is key to being an effective Agent (Brown, 2013).

Brown (2013) explains that Target characteristics include a collectivist (society or group focused) versus an individualistic society (individual focused), and whether the Target is suggestable versus stubborn. Young and old Targets tend to be the most easily influenced, that is, suggestable. These factors explain in part why it is important to know your audience.

Brown (2013) explains that while certain aspects of Context are set, others can be manipulated, especially through the principles of influence noted earlier. For example, Agents can create the perception of:

- Scarcity through statements such as 'call while supplies last'

- Authority through having an actor wear a white lab coat in a healthcare commercial

- Social proof statements such as '4 out of 5 doctors recommend'; '10 million users can't be wrong'

7.2.3 Understanding power

Power and influence are not the same thing, but they are related. People with power can make others do things, but this may not yield the desired outcomes. Moreover, individuals who do not have power can still be influential. Understanding power is key to understanding how we can influence others when we cannot force them to do what we want, or to believe in our ideas.

The following material is summarized from online material provided by the University of Minnesota (2010). Having power and using power are two different things. For example, managers (bosses) have the power to reward or punish employees. When managers make a request, it will usually be followed even without direct or immediate reward. The ability to, or the possibility of, reward and punishment is sufficient for employees to follow the request. However, there is more to power than reward and punishment, and these other aspects of power are important in scientific settings.

Researchers have identified six general sources of power listed in Table 7.2. An individual might have power from one source or all six, depending on the situation (University of Minnesota, 2010).

Legitimate power comes from one's role or position, as when a boss makes a request, a teacher assigns homework, or a police officer makes a traffic stop. Most people comply with such requests because we accept the legitimacy of the position, whether we like or agree with the request or not. CEOs or high-level executives set deadlines and employees typically comply even if they think the deadlines are too ambitious (University of Minnesota, 2010).

Reward power is the ability to grant a reward, such as a pay increase, bonus, or other perk. Reward power tends to accompany legitimate power and has the greatest influence when the reward

TABLE 7.2 Six Sources of Power

Legitimate
Reward
Coercive
Expert
Information
Referent

is scarce. However, anyone can use reward power, for example, via public praise or giving someone something in exchange for their compliance (University of Minnesota, 2010).

Coercive power is the opposite of reward power because it stems from the ability to take something away or to punish. Although the idea of power through punishment is less appealing than power via reward, in Chapter 5 on cognitive biases we saw that loss aversion motivates us more strongly than a gain of equal value. Hence, coercive power is influential. Coercive power often works through fear and forces people to do something that they would otherwise not do. The most extreme example of coercion is government dictators who threaten physical harm for noncompliance. Parents may also use coercion such as grounding their child as punishment for noncompliance (University of Minnesota, 2010).

Expert power comes from knowledge and skill. Steve Jobs had expert power from his ability to know what customers wanted even before they could articulate it. Senior employees often have expert power. Science and technology companies are often characterized by expert, rather than legitimate power. Many of these firms utilize a flat or matrix structure in which clear lines of legitimate power become blurred as everyone communicates with everyone else regardless of position. As the old saying goes, knowledge is power (University of Minnesota, 2010).

Information power is like expert power but differs in source. Expert power stems from knowledge or skill, whereas information power stems from access to specific information. Within organizations, a person's social network can either isolate them from information power or create it. There is an old saying: 'it's not what you know, it's who you know.' That saying is based on information power. However, the saying would be more accurate if it were: 'It is who you know and/or what you know.' Those who can span boundaries and connect different parts of an organization often have a great deal of information power (University of Minnesota, 2010).

Referent power stems from the personal characteristics of the person such as the degree to which we like, respect, and want to be like them. Referent power is often called charisma – the ability to attract others, win their admiration, and hold them spellbound (University of Minnesota, 2010). Examples in US political history include Abraham Lincoln, Franklin Roosevelt, Martin Luther King Jr., and John Kennedy.

7.2.4 Influence Tactics

Researchers have identified distinct influence tactics. How these tactics are applied differs in subtle ways between how bosses, subordinates, and peers use them. These differences will be discussed after first focusing on the nine influence tactics (University of Minnesota, 2010; Table 7.3).

Recall that in the ATTC model responses to influence attempts include resistance, compliance, or commitment. Resistance occurs when the influence target does not comply with the request and either passively or actively repels the influence attempt. Compliance occurs when the target is not enthusiastic about or committed to the request but does so anyway. Commitment occurs when the target agrees to the request and actively supports it (University of Minnesota, 2010).

Rational persuasion includes using facts, data, and logical arguments to convince others of a point of view. This is a tactic well-suited to statisticians (University of Minnesota, 2010).

TABLE 7.3　The Nine Influence Tactics

Rational persuasion
Inspirational appeals
Consultation
Ingratiation
Personal appeal
Exchange
Coalition
Pressure
Legitimizing

Inspirational appeals seek to tap into our values, emotions, and beliefs to gain support for a request or course of action. When President John F. Kennedy said, 'Ask not what your country can do for you, ask what you can do for your country,' he appealed to the higher selves of an entire nation. Effective inspirational appeals are authentic, personal, big-thinking, and enthusiastic (University of Minnesota, 2010).

Consultation refers to the influence agent's asking others for help in influencing or planning to influence another person or group. Consultation is most effective in organizations and cultures that value democratic decision-making (University of Minnesota, 2010).

Ingratiation refers to different forms of making others feel good about themselves. It includes flattery done either before or during the influence attempt (University of Minnesota, 2010).

Personal appeal is helping another person because you like them, and they asked for your help. We enjoy saying yes to people we know and like (University of Minnesota, 2010).

Exchange is give-and-take in which someone does something for you, and you do something for them in return (University of Minnesota, 2010).

Coalition tactics refer to a group of individuals working together toward a common goal to influence others. Common examples of coalitions within organizations are unions that may threaten to strike if their demands are not met. Coalitions also take advantage of peer pressure. The influencer tries to build a case by bringing in unseen allies to convince others. An example is someone saying I have already run this idea by person X, and she agrees with me. Another example is using client lists to promote goods or services. Knowing that a client, especially a known and/or respected client, bought from the company is a silent testimonial (University of Minnesota, 2010).

Pressure refers to exerting undue influence on someone to do what you want or else something undesirable will occur. This often includes threats and frequent interactions until the target agrees. Research shows that managers with low referent power tend to use pressure tactics more frequently than those with higher referent power. Pressure tactics are most effective when used in a crisis and when they come from someone who has the other's best interests in mind, such as getting an employee to an assistance program to deal with a substance abuse problem (University of Minnesota, 2010).

Legitimating tactics occur when the appeal is based on legitimate or position power. 'By the power vested in me ...': This tactic relies upon compliance with rules, laws, and regulations. Legitimizing does not motivate people like referent power does, but it can align them behind a direction (University of Minnesota, 2010).

The type of influence tactic that is available to any individual depends on the source of their power. Figure 7.2 illustrates these relationships.

Moving from left to right in the figure, the source of power transitions from the position held by the influencer to the persona of the influencer. With legitimate power, an influencer has at her/his disposal rational persuasion, ingratiation, consultation, and legitimizing influence tactics. Reward power supports the exchange tactic. Coercive power supports pressure. Connection power supports coalition. Information and expert power support

Power Source	Position ←————— ————→ Personal						
Types of Power	Legitimate	Reward	Coercive	Connection	Information	Expert	Referent
Influence Tactics	Legitimate	Exchange	Pressure	Coalitions	Rational Persuasion	Rational Persuasion	Inspirational Appeal
	Consultation						Personal Appeal
	Rational Persuasion						
	Ingratiation						

FIGURE 7.2 The relationship between the source of power and influence tactics.

rational persuasion. And, referent power supports inspirational and personal appeals.

The type of influence tactic also tends to vary based on the target. For example, different influence tactics may be appropriate to use with your boss than with a peer or with employees reporting to you. Peer influence is common. However, to be effective, peers should influence without being competitive. Rational persuasion is the most frequent influence tactic among peers (University of Minnesota, 2010).

Upward influence targets bosses and others in positions higher than the Agent. Upward influence may include appealing to a higher authority or citing the firm's goals as an overarching reason to follow a cause. Upward influence can also be an alliance with a higher status person. As the complexity or size of an organization grows, the benefit from upward influence increases because no one person at the top of the organization can know enough to make all the decisions. Moreover, even if someone did know enough, time would not permit making all the decisions in a timely manner. Therefore, many individuals at differing levels of the organization make or influence decisions. Therefore, statisticians even early in their career can be more effective by improving their influence ability (University of Minnesota, 2010).

By helping higher-ups to be more effective, employees can gain power for themselves and their team or function. On the flip side, leaders allowing themselves to be influenced by those lower in the organization can build their credibility and power as a leader who listens.

Downward influence is influence on employees lower in the organization or of lower status. This is often best achieved through an inspiring vision. Articulating a clear vision helps people see the end goal and move toward it. Specifying what needs to be done to get there is not necessary because people will figure that out on their own, often more effectively than the leader who is less well-versed in the nuance of required actions. Moreover, employees who participate in making the plan are more likely to

TABLE 7.4   Frequency of Influence Tactics and Their Outcome

	Frequency of Use (%)	Resistance (%)	Compliance (%)	Commitment (%)
Rational persuasion	54	47	30	23
Legitimating	13	44	56	0
Personal appeals	7	25	33	42
Exchange	7	25	33	42
Ingratiation	6	41	28	31
Pressure	6	56	41	3
Coalitions	3	53	44	3
Inspirational appeals	2	0	10	90
Consultation	2	18	27	55

Source: University of Minnesota (2010).

be committed to executing the plan, especially when the inevitable difficulties in doing important things arise (University of Minnesota, 2010).

Table 7.4 summarizes research results on how frequently each of the influence tactics is used and the outcome of the influence attempt. Rational persuasion was used most frequently. However, it resulted in half the Targets being resistant. Inspirational appeals and coalitions were not used frequently but were highly effective. Pressure, legitimating, and coalitions yielded poor results.

7.3 LEADERSHIP

7.3.1 Foundational Ideas

Brown (2013) outlines two main types of leadership, transactional and transformational. Transactional leadership is based on exchange and reward, and sometimes on pressure. Examples of transactional leadership can often be characterized by 'If you do X, then I'll do Y.' Transactional leadership can be best for day-to-day activities because workers respond well to the clarity and transparency it brings.

Transformational leadership is best for big-picture, long-term tasks because it provides the inspiration needed to endure; it changes hearts and minds. Transformational leadership can inspire positive changes in both individual employees and the entire organization. Transformational leadership is often an essential element in creating a culture of innovation and successful outcomes, as evidenced by the high rates of commitment from influence attempts based on inspirational appeal (Brown, 2013).

Transformational leadership is often equated with leaders' personality, thinking that strong, enthusiastic, and/or passionate personalities are what drive transformational thinking. However, the characteristics of transformational leadership are more complex than innate personality traits. Transformational leaders should develop behaviors, strategies, and actions grounded in leadership theory.

The four 'I's of transformational leadership (Brown, 2013) are listed in Table 7.5 and explained after the table.

7.3.1.1 Intellectual Stimulation

Transformational leaders question the status quo, challenge assumptions, and encourage the same mindset in others. Intellectual stimulation entails new experiences, new opportunities, and new ways of thinking. Emphasizing opportunities to grow and learn, rather than focusing on the outcomes of the efforts, transformational leaders reduce or remove the fear of failure, thereby empowering employees to learn and seek opportunities rather than playing it safe (Brown, 2013).

7.3.1.2 Individual Consideration

A key in transformational leadership is transmitting a sense of the larger culture to individuals, giving employees a feeling of ownership in company goals and autonomy in the workplace.

TABLE 7.5   The Elements of Transformational Leadership

Idealized influence
Inspiration
Intellectual stimulation
Individual considerations

Transformational leaders do not dictate ideas for employees to carry out, but instead focus on the professional development of employees and fostering positive relationships with them (Brown, 2013).

Transformational leaders are often easy to identify by the trust, respect, and admiration others feel for them. Transformational leaders do not micromanage. They lead by communicating a clear vision and creating a workplace where employees are trusted to make decisions. Employees are encouraged to find new solutions to longstanding challenges (Brown, 2013).

7.3.1.3 Inspiration

Employees want leaders to impart a vision that is appealing and worthy. This requires communicating a vision that followers internalize and make their own. To accomplish this, employees benefit from a strong sense of purpose with high standards and expectations for achievement (Brown, 2013). The motivation to achieve cannot be fear-based. It must be inspired by example. Transformational leaders set high standards and expectations for themselves and then model them for their employees. Leaders' actions instill passion into their followers (Brown, 2013).

7.3.1.4 Idealized Influence

Transformational leaders serve as role models for employees. This includes modeling ethical and socially desirable behavior, maintaining a dedication to goals, and exhibiting enthusiasm about company strategy (Brown, 2013).

The foundation of idealized influence is trust and respect. Leaders who have developed idealized influence are trusted and respected to make good decisions for the organization, and for the individual teams and employees within it. With this trust, others want to emulate their leaders and internalize their ideals (Brown, 2013).

Transactional and transformational leadership take differing approaches to motivation. Transformational leadership is for long-term goals, with focus on individual and organizational growth instead of short-term achievement. Transactional leaders establish criteria for success and then reward or penalize team members based on how well they perform. Transactional leadership is results oriented and tends to be more appropriate when the goal is to complete specific tasks within a limited time frame.

Both styles of leadership can be utilized and even combined to best serve a team's function and achieve desired goals. It falls to the leader to recognize which leadership approach can best motivate employees to achieve specific goals.

7.3.2 Perspective on Leadership

Doris Kearns-Goodwin is a noted presidential historian. Her book on Abraham Lincoln, *Team of Rivals* (Kearns-Goodwin, 2006), was made into a well-known movie. Her books on Teddy Roosevelt, Franklin Roosevelt, and Lyndon Johnson are also compelling. Her book *Leadership* (Kearns-Goodwin, 2018) is a masterful illustration of leadership in action. Key points on leadership are illustrated through the lives of the four presidents she spent decades researching. Kearns-Goodwin notes that these four presidents were in some ways very different. They were from different areas of the country, coming from both humble and privileged backgrounds. Lincoln showed political promise from an early age, while Roosevelt (FDR) was a relative latecomer to politics. However, each led the country during difficult periods of change and upheaval.

In *Leadership*, Kearns-Goodwin (2018) observed that there's 'no master recipe for leadership', she did note 'a family resemblance of leadership traits.' Aside from the key attributes of humility, empathy, resilience, and the ability to control negative impulses, here are a few of the common threads she identified.

7.3.2.1 Leaders Can Be Made

According to Goodwin, the four presidents had different degrees of innate leadership potential. But, she says, 'Far more important than their inborn qualities was the work ethos that they developed. The real success that most people have is when their ordinary talents are developed to an extraordinary degree through focused, sustained work'.

7.3.2.2 Communicate through Stories, Not Just Facts and Figures

In the section on cognitive biases in Chapter 5 on critical thinking and making decisions under uncertainty, it was noted that people tend to remember stories better than data. Hence, communicating through anecdotes can be helpful in building common ground. Lincoln's speeches always told a story. Theodore Roosevelt excelled at short, punchy statements. Lincoln believed people remember stories because they have a beginning, middle, and end – and the sequence sticks with them.

7.3.2.3 Listen to a Variety of Opinions

Kearns-Goodwin notes the importance of surrounding ourselves with people who are willing to disagree without fear of consequence. This ties into ideas on well-functioning groups discussed in Chapter 6. Lincoln included his political rivals Edward Bates, Salmon P. Chase, and William H. Seward in his cabinet because he valued the diversity of their opinions and advice. This is a valuable lesson for leaders at every level in every discipline.

7.3.2.4 Resilience

Leadership studies, Kearns-Goodwin says, suggest that resilience, the ability to sustain ambition through loss and adversity, is at the heart of leadership growth. Each of the presidents in Goodwin's book suffered devastating losses and was subsequently able to grow through their experiences. Being able to admit mistakes and reframe setbacks as growth opportunities is an invaluable skill.

7.3.2.5 Make Time to Relax and Reflect

The ability to find time to think is crucial, Goodwin says, and an 'underappreciated' attribute. It is important to take time to relax, to replenish energy, and to find balance, which is harder than ever in today's world where we always carry instruments of communication. Lincoln went to the theater more than 100 times during the Civil War. When the lights went down and a Shakespeare play came on, 'for a few precious moments, he could forget the war that was raging.' Similar ideas were covered in Chapter 2 on Deep Work, Hyperfocus, and Scatterfocus.

7.3.2.6 Real Change Happens When Leaders and Followers Form a Partnership

'When Lincoln was called a liberator, he said, 'Don't call me a liberator. It was the anti-slavery people that did it all,' Goodwin says. Lincoln is further quoted to say: 'When the Revolution was won, it was won, people said, in the hearts and minds of the people before the first battle was even fought.'

Kearns-Goodwin's greater point is that power has to have a purpose. Most of the large-scale change in the United States has taken place when citizens and leaders work toward a common goal. Leaders should gain popular support and work with those they are leading, or they will have difficulty accomplishing their goals. Describing the point in this manner makes clear the connections with influence noted earlier in this chapter. Change comes from commitment and an essential component of commitment is inspirational leadership.

7.4 PUTTING THE PRINCIPLES INTO PRACTICE

Influence is an inherent part of leadership. However, effective influence can be applied by good Agents and bad Agents. Therefore, being a good leader entails being a good Agent. Therefore, it is useful for leaders to study the actions of good and bad Agents. Many bad Agents did not start with bad intent, but somehow went astray. Unfortunately, we have numerous examples from which to pick for examination.

Aspiring leaders should take advantage of the many team and functional meetings that provide opportunities to practice the influence skills outlined in this chapter. By focusing on these opportunities and soliciting feedback, influence and leadership skills can be cultivated, thereby preparing aspiring leaders for larger-scale opportunities for influence and leadership later.

In *Making Critical Decisions* (Roberto, 2013), the author Michael Roberto notes that effective leaders should be problem finders, not just problem solvers. Organizational breakdowns tend to evolve over time, beginning with small errors that are compounded and eventually gain momentum. The best way to prevent big failures is to catch and correct small errors as soon as possible. To do this, leaders cannot think they have all the answers. Instead, leaders can focus on shaping and directing an effective decision-making process, marshaling the collective intellect of those around them.

Leaders should focus on process (e.g., decision-making process), not just content (e.g., data used to inform the decision). Noted business consultant and author Peter Drucker is quoted in Roberto (2013) as saying, 'The most common source of mistakes in management decisions is the emphasis on finding the right answer rather than the right question.'

In situations where intuitive thinking can successfully drive decision-making, it is important for leaders to explain their intuition and the rationale behind their decisions so that others understand and trust the decisions. Decisions based on intuition may require more effort to build the trust in employees needed for successful implementation because employees will not share the leader's intuition (Roberto, 2013).

Other specific things good leaders should focus on include (Roberto, 2013):

- Provide a process road map at the outset of the decision process

- Reinforce an open mind-set

- Engage in active listening

- Explain decision rationale

- Explain how others' inputs were employed

- Express appreciation for everyone's input

An important part of good leadership stems from the ideas on critical thinking and decisions under uncertainty discussed in Chapters 5 and 6. Leaders and senior team members should promote an organizational environment that allows teams to function well. This includes avoiding the cultures of yes, no, and maybe. It also entails avoiding the five dysfunctions of teams (see Chapter 6) and promoting an environment for individuals to do Deep Work (see Chapter 2), to be creative (see Chapter 3), and to think critically and make decisions under uncertainty by putting systems in place that avoid cognitive biases (see Chapter 5).

7.5 REAL-WORLD EXAMPLES

Although leadership can be different things in differing situations, the following two quotes illustrate two key aspects of leadership.

- 'Your most important task as a leader is to teach people how to think and ask the right questions so that the world does not go to hell if you take a day off.' Jeffrey Pfeffer

- 'Leaders should influence others in such a way that it builds people up, encourages and edifies them so they can duplicate this attitude in others.' Bob Goshen

Leadership skills can be improved outside the workplace and within the workplace in ancillary roles. For example, I coached youth athletics and led a volunteer organization. These experiences were useful in enhancing my leadership skills. Also, when I first joined the pharmaceutical industry, my company had a one-day conference for statisticians each year. New employees organized the event and were also encouraged to organize a therapeutic area seminar series. These were useful experiences in developing leadership skills – and helped develop an internal network for the new employees.

Figure 7.3 shows the diagram, originally created by Steve Ruberg and passed to me while he and I were colleagues at Eli Lilly, is valuable, practical advice on leading innovation or complex change. The top row of the diagram illustrates that important change requires vision, skills, incentives, resources, and an action plan. Each subsequent row illustrates outcomes when one of the necessary elements is missing. Without adequate communication of each of these elements, the outcome will be the same as if the element is missing.

An inescapable fact from reviewing the figure is that no one person can provide all the things necessary to lead and effect disruptive and/or complex change. A skillful leader can provide the vision and influence others through referent power. If that visionary leader is a high-ranking executive, she/he may be able to also provide the incentives and resources. However, it is unlikely that this same person can teach others all the skills

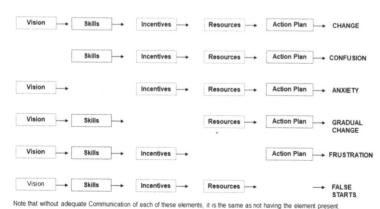

FIGURE 7.3 Diagram of the necessary elements for leading innovation or complex change.

needed, nor is it likely that such a high-ranking executive would have sufficient knowledge of the organization and foundational skills to develop an action plan, or to communicate all aspects of the change initiative. Therefore, great leadership involves putting together a great supporting cast so that all the needed pieces can be put into place.

The prospect of leading a group, even a small group, can be intimidating. Many potential leaders worry that they lack the inherent skill required to lead. Hopefully, this chapter helps you to see that although some have greater inherent leadership abilities than others, much of what good leaders do can be learned. The most important point on leadership I have ever encountered is that if you care about the people you lead, you can be a good leader using any number of leadership styles and a variety of skill sets. But, if you do not care about the people you lead, you are unlikely to be a good leader no matter what style you use or how smart you are.

CHAPTER 8

Effective Work
Relationships

ABSTRACT

This chapter covers relationships in the workplace and why they are so important to our success and happiness. The chapter begins with a section on how to create a healthy work environment for ourselves. At first, it may seem counterintuitive to focus on self when trying to work well with others. However, a key point of the chapter is that being well-grounded and happy ourselves is foundational to successful relationships with others. The next chapter covers foundational concepts for establishing effective working relationships with others, followed by how to give and receive feedback, and then has a section on impression management. The chapter concludes with practical advice on how to put the principles into practice, along with real-world experiences of the author that reinforce these points.

DOI: 10.1201/9781003334286-10

8.1 INTRODUCTION

The benefits of getting along with co-workers and the consequences of not getting along are substantial. Good working relationships help us and our teams to be more effective and promote greater satisfaction and happiness. Not getting along does the opposite.

According to Fisher and Phillips (2021), authentic connection and meaningful workplace relationships foster psychological well-being, but meaningful relationships at work have become more difficult. The relentless striving to increase productivity and incorporate new technologies creates less humane work environments. Recall that in Chapter 2, Deep Work and Hyperfocus were touted as means to increase productivity. It is again important to emphasize that the goal is not to increase the amount of work done, but rather to increase the efficiency and value of the work, thereby being productive while saving time for work–life balance – which includes meaningful work relationships. In the following section that has advice from Fisher and Phillips (2021) on work relationships, notice the parallels with ideas presented in Chapter 2 on productivity, prioritization, and work–life balance.

8.2 PRINCIPLES IN FORMING MEANINGFUL AND EFFECTIVE WORK RELATIONSHIPS

8.2.1 Establishing a Healthy Work Environment

It is important to appreciate that good relationships do not just happen. Loneliness is endemic in the United States (and elsewhere). Over 40% of Americans feel lonely, even when surrounded by co-workers. Only 20% forge meaningful workplace friendships; many of the others suffer disconnection, disengagement, and dissatisfaction. Loneliness increases susceptibility to cognitive decline and physical ailments. The increase in remote and gig work exacerbates this problem by reducing in-person interactions. Burnout is on the rise (Fisher & Phillips, 2021).

Moreover, being digitally connected 24/7 and deluged with information and communication, paradoxically, increases alienation. The 'Workism' lifestyle of always being online and always having at least part of our mental energy devoted to work is not sustainable long-term. Chronic overwork decreases output and causes stress-related problems such as burnout, fatigue, anxiety, depression, heart disease, sleep disorders, and strained relationships at home (Fisher & Phillips, 2021).

Strong social bonds are important in combating these workplace trends. Spending too much time in cyberspace depletes two of your most valuable resources: Time and attention. The reduction in face-to-face interactions dehumanizes work environments and threatens the psychological safety we need to share ideas, be creative, and take risks. Workism causes us to associate value with busyness. Responding to the constant assault of digital communications lessens our ability to differentiate between important and unimportant tasks (Fisher & Phillips, 2021). It is important here to make a connection. The increased efficiency and value of Deep Work (versus Shallow Work) discussed in Chapter 2 also fosters the time and energy for the meaningful relationships touted by Fisher and Phillips (2021).

Two-thirds of employees suffer burnout. The costs of burnout are numerous, including employee disengagement, decreased performance, and increased turnover. The antidote is social connection; people who report work satisfaction have strong work relationships. They suffer less stress, enjoy better attendance, and are more engaged, productive, and resilient (Fisher & Phillips, 2021).

So, work relationships are important, but how do we go about having good relationships? A good starting place is to consider work styles. Work style is a combination of character, values, skills, and proficiencies. Fisher and Phillips (2021) provide the following examples of work styles. Ethan's team works well together in a company that provides high-tech solutions to manufacturers. Ethan has a collaborative work style and excels at bringing people together to resolve issues. Nichole is meticulous, an asset for meeting deadlines and staying on budget. Chloe excels at big-picture thinking, while Josh's strength lies in his technical proficiency.

Business Chemistry is an analytical tool to assess work styles created by Deloitte that categorizes people into four personality types. 'Pioneers,' like Chloe, are creative, big-picture thinkers. Nichole is a 'Guardian': Her strength lies in nailing down the details and conforming to processes. Josh, quantitative and analytical, is a 'Driver'; diplomatic Ethan is an 'Integrator.' No one, of course, fulfills only one role exclusively; everyone operates somewhere on a behavioral continuum depending on the situation and temperament. Understanding tendencies and preferences helps us design a work life in which we are using our talents and skills at their best and allows for better connections with others (Fisher & Phillips, 2021).

When people feel comfortable, connected, valued, and valuable at work, business prospers. Connection, comfort, and contribution promote feelings of belonging and result from strong interpersonal relationships. Unfortunately, many team leaders and companies value these qualities less than productivity and performance. When well-being and healthy relationships are secondary to productivity, it's like a sports team of talented players who don't stay healthy and don't work together. Companies and teams that prioritize well-being and healthy relationships create the strongest, healthiest team dynamics. Workplaces that emphasize transactional, business-first relationships may foster a culture that thinks in terms of 'winners' versus 'losers.' Team members who work independently can miss opportunities to collaborate, pool resources, or solicit help from each other. In unhealthy organizations that do not value well-being or relationships, teams devolve into dysfunction, confusion, apathy, and low performance (Fisher & Phillips, 2021).

Hence, an important aspect of having effective and healthy work relationships is to work in an environment that supports these goals. Of course, few of us can make job choices on this criterion alone. Nevertheless, bearing in mind this important criterion can lead to better choices in where and with whom we work.

Psychological safety, empathy, and trust are the foundational pillars of a healthy work environment or team. Approaches to building a healthy team vary depending on the work style and temperament of its individual members. Psychological safety empowers team members to work through conflicts rather than avoiding conflict and to share ideas rather than hoarding them. Empathy helps employees identify and relate to each other's feelings, motives, and behaviors. Trust builds confidence that colleagues are honest, reliable, responsible, and determined – people we can feel physically and emotionally safe around. In contrast, the outdated ideas about the hardest workers get ahead contribute to unhealthy workplaces and to the poor health of individual workers. If we think working harder is the key to success, we are likely to fail, and to be miserable at work and at home (Fisher & Phillips, 2021).

To avoid the destructive Workism mindset, Fisher and Phillips (2021) advise focusing on the items listed in Table 8.1.

It is easier to implement these ideas in some environments than in others. However, we can always do things that will help. The

TABLE 8.1   Keys to Establishing Your Own Healthy Work Environment

Well-being is more than physical health. It is mental and physical welfare, including feeling secure, accomplished, and personally satisfied.

Money does not define success.

Rest is not a sign of weakness. Taking time to recharge boosts productivity and resilience, and guards against burnout.

Burnout does not go hand-in-hand with success. The human cost of working long hours and high stress is severe. Success at work does not require sacrificing your personal life. Loving your job shouldn't mean sacrificing everything else in your life.

The Workism culture is powerful and we have to actively work against falling into it.

Create a personalized well-being plan to manage career pressures.

things Fisher and Phillips (2021) recommend to include in our personal well-being plans have a strong overlap with the items noted in Chapter 2 on productivity and prioritization: Prioritize tasks, schedule time for uninterrupted work, take regular breaks during the day, take longer time away from work to recharge. Other items they advocate include a healthy diet, regular exercise, spending time with family, and maintaining a mindfulness practice.

An important item on our well-being checklist is setting boundaries on the use of technology/social media. As described in Chapter 2, social media, news feeds, etc., are designed to keep our attention, to distract us, and they will hijack our time and attention unless we actively work to combat it. Therefore, to maintain well-being, set boundaries and make intentional choices when using technology. Become aware of why, when, and how you engage with technology. Note compulsive desires to check your phone or email; make deliberate choices about how to use your time. Some useful suggestions include: pick a day each week to stay off social media, ban smartphones from the dinner table, delete apps you seldom use, and schedule device-free time.

8.2.2 Working with Others

In considering how to work well with others, it is important to realize that people skills are important, and that technology cannot replace people skills such as creativity, empathy, adaptability, and emotional intelligence. No matter how technical the job, a combination of emotional intelligence and cognitive skills is beneficial for gaining a clear understanding of your professional persona. As discussed in the section on real-world examples, I had an eye-opening experience when receiving feedback on how others perceived me at work.

Having a healthy individual approach to work and the workplace, combined with a good understanding of how you are perceived, sets the stage for maintaining or improving working relationships. My wife, Donna, is one of the most socially skilled people I know. Many of the tips given in Table 8.2 come from her,

TABLE 8.2   Methods for Establishing Effective Working Relationships

Initiate conversations by asking questions
Say thank you
Be positive
Introduce yourself and others
Let others know who you are
Share information
Support the work of others
Ask others to get involved in your work
Initiate repeated interactions
Participate with others in non-work activities

with others being items I've seen to be effective. Details on each item are provided after the table.

8.2.2.1 Initiate Conversations by Asking Questions

Meeting someone new can be intimidating because we may not know what to say or how to say it. Asking questions is a great way to listen and let the other person share. They will feel closer to you when they have shared, and you have demonstrated you are interested in what they have to say. Then share something about yourself to make the relationship a two-way interaction to help form a bond.

8.2.2.2 Say Thank You

Verbal or written signs of appreciation for those who have helped you, or to those doing exemplary work, making positive contributions, and going above the call of duty will help forge bonds. These notes can be hard written, sent via email, done by phone or voice mail, and perhaps best of all, when possible, done in person. Signs of appreciation can be given to people above you, below you, or at the peer level. Colleagues like to be appreciated and will feel closer to you by having been noticed and thanked for their contributions. When thanking a peer or someone below you, consider including their manager so that the recognition is shared.

8.2.2.3 Be Positive

Grumpy complainers are not well-liked because complaining brings people down. Remember the adage, if you don't have something good to say, don't say anything. The context here is not formal, private feedback sessions, but rather general workplace conversations. In those casual settings, it is best to be silent or to speak positively about the people you work with. Shared information, positive or negative, often comes back to the person being discussed. People enjoy hearing that you have said supportive things about them and will know that you are on their side. That builds trust. Do not contribute to workplace gossip. It is so prevalent that it can go unnoticed, until someone becomes aware you said something untrue or out of context about them.

8.2.2.4 Introduce Yourself and Others

If you can reach out and introduce yourself to some of the people you work with or want to know better, you will find they are more inclined to let down their guard. It will be easier for you to get to know them and share who you are. Building positive relationships often provides increased resources to help you get your job done and to be more efficient. You will enjoy greater satisfaction at work, and so will those around you. Much the same will be true when you introduce someone else to a person you already know.

8.2.2.5 Let Others Know Who You Are

Although you do not want to be the one who bores others by dominating a conversation or bragging about yourself, one of the best ways to build relationships is to let others know who you are. You can achieve this by sharing your expertise, knowledge, and personality in meetings or as appropriate in other workplace and social interactions. This helps others to get to know you and to see you as more approachable, which is a building block for relationships.

8.2.2.6 Share Information

Information you share with others can be directly related to their work, or it can be about a subject they enjoy. In sharing with someone, you show that you are thinking of them and helping them by providing information, content, or context.

8.2.2.7 Support the Work of Others

Ask others how you can help. They will appreciate the help and this will help form a closer connection because you are working directly with others to help them meet their goals.

8.2.2.8 Ask Others to Get Involved in Your Work

Ask others for help. You do not have to achieve something on your own to be valuable. Moreover, bringing others into your work can provide them with valuable experience and opportunities, and they will get to know you better.

8.2.2.9 Initiate Repeated Interactions

Continued interaction is an important part of building relationships with people you have met. A one-time meeting is not a relationship, it is a start. Through repeated interaction, you get to know each other better and establish a closer connection.

8.2.2.10 Participate with Others in Non-work Activities

Meet with others for lunch, go to the gym, go for a run. As you get to know someone a little bit, you may find similar interests that warrant activity outside of work and begin the process towards friendship.

8.2.3 Giving and Receiving Feedback

Giving and receiving feedback is difficult but important. We will improve more, help others improve more, and get along better with co-workers if we make use of and give constructive feedback and advice. The points in Table 8.3 and the explanations after the

TABLE 8.3   Tips on Giving and Receiving Feedback
- Be professional, not personal
- It's a two-way conversation
- Focus on facts rather than feelings
- Be direct
- Balance positives and negatives
- Choose words carefully
- Focus on fixing not finding fault

table are summary notes from a training course that I took on giving and receiving feedback.

8.2.3.1 Be Professional, Not Personal

At times, everyone will have a strong emotional response to an employee error or to the comments from a supervisor. Being upset is understandable, but it will not fix the situation. As a supervisor, check your temper and wait until you are calm enough to deliver measured feedback. As an employee, if you receive unexpected negative feedback listen to the feedback and ask for time to process the information and continue the conversation later. Speaking when emotions are high is unlikely to help, whether we are giving or receiving feedback.

Supervisors cannot expect employees to be open to criticism when the employee is immediately put on the defensive by blaming or shaming. And employees cannot expect supervisors to be sympathetic to their side of the story when they till it in the heat of anger or frustration. Time is our friend. Use it. If the feedback is important, if your side of the story is important, it probably can wait until you count to 10 – it can wait a day – until you can communicate more effectively. These conversations should always be in private.

8.2.3.2 It's a Two-Way Conversation

Whether giving or receiving feedback, keep an open mind and give the other person opportunity to explain their side of the story. Employees will often admit to shortcomings, ask for help, or explain legitimate extenuating circumstances when given the chance. Supervisors will often be able to put the feedback in the most meaningful context for an employee if the employee is allowed to explain. By letting the other person be heard, you may realize that the problem is a symptom of a larger underlying issue.

8.2.3.3 Focus on Facts Rather Than Feelings

It is important to address the specific problem, not your frustrations. For example, if an employee missed several deadlines, instead of a reprimand such as 'I'm tired of you missing deadlines!' spell out how the person's actions impact the team. You might say, 'When you're slow to complete your portion of a project everyone is affected because we all have to stay late to meet our obligations.' Then, offer specific suggestions to help the individual solve the problem. If a supervisor reprimands you for something that is not your fault, do not say: 'I'm tired of being blamed for things that are not my fault!' Instead, provide the context the supervisor needs to understand the situation.

8.2.3.4 Be Direct

Sweeping problems under the rug is an easy way to avoid difficult conversations. Mincing words and giving half-truths can make the conversation easier. But withholding negative feedback is a disservice to underperforming employees because it deprives them of information they could use to improve. Be kind but candid. Say what needs to be said in a tactful yet straightforward way. Instead of vague language, be clear. As an employee, remember that a supervisor cannot fix the problem if you are not candid. For example, if a family member has a health problem, you do not need to tell your supervisor the specifics. However, it may be better to say you have been distracted and troubled by the illness of a loved one, rather than just saying things are a little difficult outside of work. This will help the supervisor understand the gravity of the situation. These are good examples of how communication can be more effective if done when emotions are not overriding the conversation.

8.2.3.5 Balance Positives and Negatives

For supervisors, knowing how and when to provide criticism is an important skill, but do not provide feedback only when employees slip up. Complementing well-done work and recognizing improvements is an excellent way to boost morale, reinforce positive behavior, and allow employees to be more receptive to negative feedback. As an employee, it can be useful to acknowledge that feedback in the context of other tasks was done well. But be careful here. This can seem like dodging negative feedback. However, if appropriate, you may need to tactfully remind your supervisor that although you were late or did not perform as well as hoped on project Z, projects X and Y were higher priorities. Again, be very careful here and do not push back when emotional.

8.2.3.6 Choose Words Carefully

Most employees know when they make a major mistake and do not need help feeling embarrassed. Avoid demoralizing statements that call into question the employee's intelligence. Also, avoid subjective statements such as 'You are not motivated enough'; and avoid generalizations such as 'You never contribute new ideas during meetings.' As an employee does not respond with subjective statements such as 'You never give me enough credit when I do well'; and avoid general responses such as 'You never listen to my side of the story.'

8.2.3.7 Focus on Fixing the Problem

Whenever giving feedback, remember the goal: To rectify the issue. Whether you have to provide the employee with additional training, offer more frequent direction, or streamline a flawed system, do what you can to help the employee correct the problem. As an employee, do not just point out the problem that resulted in getting the feedback. If possible, offer potential solutions and your willingness to help implement them.

8.3 IMPRESSION MANAGEMENT

Impression management can be a touchy topic for scientists, especially introverts. Impression management can seem like 'brown-nosing' or 'sucking up.' When done properly, impression management is helping others to see who you really are, not the slimy act of a corporate climbing, fake-it-till-you-make-it imposter.

Common methods of impression management include the choice of clothing, the avatars or photos used to represent ourselves online, descriptions in résumés and online profiles, and how we comport ourselves in the workplace. Appropriate impression management builds credibility, maintains authenticity, is believed by others, and fosters a variety of favorable outcomes for the individual, his/her team, and the organization.

Not all aspects of our true selves must be disclosed in the workplace. However, trying to win social approval by suppressing too much of one's true self can lead to psychological distress and unfavorable outcomes. It is important to recognize that whether we manage our professional image or not, co-workers are forming impressions. They watch our behavior and draw conclusions about the kind of person we are, whether we are trustworthy, whether we are dedicated to team goals, and how we will react in difficult situations. As my father often reminded me, 'Many people can do good or look good when things are going good. It's when things are not going good that you see someone's true character.'

Because others form impressions about us regardless of whether we try to influence those impressions, it is important to take charge and help them arrive at an impression that is both true and useful. To do this, ask yourself how you want to be seen. What qualities or character traits do you want to convey? Then, ask yourself what the professional expectations are of you and what aspects of your social identity you want to emphasize or minimize in interactions with co-workers. If you want to be seen as a leader, you might disclose how you organized an event or led a volunteer organization. If you want to be seen as a caring person in whom people can confide, you might disclose that you volunteer at a crisis helpline. You can use a variety of impression management strategies to accomplish the outcomes you want. The following are the general categories of strategies for impression management.

Nonverbal impression management includes clothes, demeanor, and posture. Once, I was interviewing a statistician for a lower to mid-level position at a large pharma company. The statistician was a recent graduate from a prestigious university. Shortly after the interview began, the statistician leaned back in his chair and propped his feet up on my desk. His posture sent a clear signal of disrespect and disinterest. It was a short interview with an easy decision.

Another example of nonverbal impression management that has received increasing attention recently is body art. Although the number of people with tattoos and piercings has increased in recent decades, body art may create unfavorable impressions in initial job interviews and interactions.

Verbal impression management includes tone of voice, rate of speech, what we choose to say, and how we say it. Managing how we project ourselves can alter the impression that others have of us. For example, excessive use of slang or colloquialisms can send the impression of being unsophisticated or lacking education.

The use of swear words is another tricky topic. Swear words are sometimes used for emphasis or to show that we are casual and comfortable. Perhaps swear words can be effective in achieving these goals; however, it is risky, and some swear words are riskier than others. I used to joke that I was allowed to use one swear word per day. However, after working with several high-ranking scientists and executives who used swear words regularly, I saw that swear words were ineffective and often did more harm in communicating than good. I now believe that in the workplace if I use one swear word per day, then that is one too many. If swearing is part of your authentic self, remember that you do not have to reveal your entire self in the workplace.

Behavior impression management includes how you perform on the job and how you interact with others. Complimenting your boss is an example of impression management. Other impression management behaviors include conforming, making excuses, apologizing, promoting your skills, doing favors, and making desirable associations known. Impression management is related to higher performance ratings by increasing liking, perceived similarity, and network centrality. However, incessant bragging and name-dropping are very different and will not foster liking.

8.4 PUTTING THE PRINCIPLES INTO PRACTICE

A quote that helps put the challenges of workplace relationships in perspective is the following:

> Practice makes permanent.

The more recognizable quote is practice makes perfect, but that is only true when the practice is of high quality. Practice engrains whatever we do as permanent, whether the practice is good or bad, precise or sloppy. This holds true for unconscious practice as well. Often, we do not consider our activities as practice, but repetition will engrain the behavior, whether conscious or not.

Therefore,.we should be conscious of our actions regarding relationships. It is easy to fall into routine workplace gossip and negativity unless we work to avoid it. Knowing the principles of workplace relationships won't make any difference unless we practice them with intent and purpose.

For example, remind yourself on your way into work, or as you log on to a work meeting, to say positive things, to compliment others, to look and act like the person you want to be. If you do that, others will see you for who you really are, and that is not brown-nosing or sucking up.

8.5 REAL-WORLD EXAMPLES

Although technical challenges abound in statistics and statistical careers, it is not unusual for interpersonal relationships to be more troublesome than technical challenges. In my own case, I cannot remember many (or any?) sleepless nights caused by technical challenges. But I did lose more sleep than I care to admit over relationships and personnel matters.

Early in my career another statistician and I had joint statistical leadership responsibilities on a team researching a drug for two indications (diseases). I was lead for indication A and the other statistician, I will call Yolanda, was head of indication B. When Yolanda asked and was allowed to lead a new study in indication A, I took that as an affront to my leadership. Instead of working out the situation with our supervisor, our relationship soured. We did not trust each other, and both of our performances suffered. Although the supervisor should have proactively discussed and explained his decision to me, when he did not, I should have sought out the supervisor for help in understanding and resolving the matter.

On another team later in my career, another statistician, I will call Jane, and I again had joint leadership responsibility. Jane and I had not previously worked on the same team, but we did work in the same therapeutic. We collaborated effectively when in a prior role Jane sought my consultation on methodological issues. We worked out a unique way of splitting statistical responsibility on our team. Jane was an expert in getting work done. She was excellent in all operational aspects of data management, statistical reporting, and developing younger statisticians in these areas. I focused on methodology and strategy. We supported each other and had each other's back. When we had different perspectives on an issue, we worked through those differences behind the scenes and presented a united statistical front to the rest of the team. It was an effective, symbiotic relationship based on trust that helped our team get a drug approved for three indications in a therapeutic area that was new to the company.

I have established several productive collaborative relationships with statisticians that do not work in my company. These relationships began because a mutual friend made the initial introduction. These relationships grew because we followed up on the initial introduction and because we found common interests outside of work.

An area where I received some surprising but useful feedback was regarding impression management. Well into my career, the VP in my department told me that some people found me to be cold and a bit standoffish. Although I am not the life of any party, I had never had feedback like that. Upon further discussion, the VP stated that he was told I was unfriendly when passing in the hallways. For context, this was a large workplace, with thousands of employees in numerous interconnected buildings that cover several city blocks.

Due to my vision impairment, I often missed subtle greetings from passersby in the hallways, such as a smile or small hand wave. I could not see their faces well enough to recognize them, nor did I realize they were trying to get my attention. I needed to be more forthright about my impairment. This was difficult for me because as a youngster I was teased and bullied because of my coke-bottle thick glasses and all I wanted at that time was to fit in and not be noticed. I was hesitant to tell people about my impairment at work, I suppose, because I was still self-conscious about fitting in. Later in life, the external evidence of my impairment was less noticeable because I wore contact lenses. I was having trouble fitting in because I did not disclose my impairment – because I was worried that I would not fit in if I disclosed my impairment. With the wisdom of hindsight, I would have been better off, I would have managed my impression better, if I had been more forthright.

III

Planning and Growing

Career Planning and Continued Learning

ABSTRACT

Chapter 9 begins Part 3, planning and growing. The chapter begins with considerations for careers and career progression, including dimensions of career planning and progression, such as duration of planning cycle, depth and breadth of expertise, theoretical and applied expertise, academic and industry orientation, managerial and technical orientation, and size of organization. The chapter continues with a section on coaches, mentors, coaching, and mentoring, and then a section on continued learning. The importance of being able to learn from routine experiences on the job is emphasized. The chapter concludes with practical advice for putting the principles into practice along with real-world experiences of the author. A key theme of the chapter is that while career planning is important, we need to appreciate that for many of us, much of what happens in our careers will not be closely tied to some grand plan, many things are unpredictable, and we need to be ready for, and to embrace, this element of our careers.

DOI: 10.1201/9781003334286-12

9.1 INTRODUCTION

Most of the material in this chapter reflects my experiences and ideas, although similar ideas can be found in many places. Because careers in statistics can be so diverse, the focus here is on general ideas rather than specific advice.

Careers are a journey for which it is hard to know the destination. Although some know early on what they want to do, many of them will eventually set their sights on a new destination. A common piece of advice is to follow your passion. This advice is sound because success requires focused work, and focused work is easier to sustain when motivated and passionate. However, this advice should also come with a significant caveat.

How can you be sure about your passion early in life or early in your career? I worry that we often chase our first passion, not our greatest passion. Therefore, even after identifying a job, it can be useful to continue exploring related and tangential options that in some way tap into our skills and interests.

The preface to this book includes the following passage: In a job interview, a hiring manager told a prospective employee: You have the perfect background for this job; the other 90% of what you need to know you can learn on the job. Although the exact percentages may be an exaggeration for emphasis, we need to continue developing and learning. But there is so much to learn, how do we go about it? How do we organize a series of experiences and roles to move toward our destination? Oh, and what is our destination? Where do we want to go in our career?

Careers are so different, specific advice here would not be useful. However, the following general concepts, or dimensions in career progression, may provide useful guidance.

9.2 CONSIDERATIONS IN CAREERS AND CAREER PROGRESSION

Table 9.1 summarizes some of the key dimensions in career planning and career progression, with explanations following the table.

TABLE 9.1   Dimensions in Career Planning and Career Progression

Duration of planning cycle
Depth and breadth of expertise
Theoretical and applied expertise
Academic and industry orientation
Managerial and technical orientation
Size of organization

9.2.1 Duration of Planning Cycle

In most organizations, work planning for individuals is tied to the calendar and it focuses on the tasks to be performed in that year. These yearly performance management plans are necessary, but they are not a substitute for career planning. Longer-term plans are also needed, for example, planning over a three-to-five-year duration. Look ahead to a possible next role and consider what skills and experiences are required for that role. Make plans, at least general plans, for how those skills and experiences can be acquired as you prepare for that role.

Examples of these mid-duration plans include:

- Working in a supportive role prior to taking on a lead statistician role for a certain type of task or project

- Collaborating on research outside of current core expertise

- Supervising a small group of workers prior to a broader managerial role

- Serving on a project or protocol review committee before taking on a broader methodological role

- Leading a youth sports team or volunteer organization to prepare for an initial leadership role at work

In addition to these mid-duration plans, more general long-term plans can be useful. Developing long-term career plans requires gathering information about the possibilities, which can be done through exposure to a variety of roles and experiences and through consultations with more senior employees. We should seek out those things we cannot see from our current role 'so we know what else is out there.'

9.2.2 Depth and Breadth of Expertise

Statisticians can make valuable contributions by either having deep expertise in one area or by having broad expertise across areas. Over a long career, it is probably necessary to have expertise in more than one area, even if one area is more notable than the others. The benefits from broader exposure were discussed in Chapter 3 on innovation and creativity and in Chapter 5 on critical thinking and making decisions under uncertainty.

Early in a career, it can be useful to first develop depth in a narrow area. This allows us to be recognized as an expert in that area, which can provide initial traction in career progression. However, too narrow a focus for too long can foster a lack of understanding of the big picture and limit our ability to recognize other areas where we can contribute, which will limit career development in the long run. Therefore, the mid-duration plans noted above should include some diversity in experience and exposure to different areas and ideas.

9.2.3 Theoretical, Applied, and Bridging Expertise

Statisticians are sometimes categorized as being 'theoretical' or 'applied.' The context in these categorizations is often related to whether the statistician focuses on researching new methods, often in an academic setting, or applying existing methods, often in an industry setting. This, like many categorization schemes, may be an over-simplification but useful, nonetheless.

Without theoreticians, applied statisticians would not have many methods to apply; and, without applied statisticians, there would be little need for new theory and methods. This mutual need for theoretical and applied statisticians leads to a third categorization that I think of as bridging expertise: Those who bridge the gap between theory and application. Bridgers have sound theoretical backgrounds and understand complex theory and new methods. However, bridgers also understand applications and see areas where the new theory and methods, or novel applications of older methods, could be useful.

The main point here is that the world needs many types of statisticians with diverse expertise and experience. It is hard to know early in our career where we fit best. Hence, there is a need, as noted in the previous section, to plan to gain exposure and experience. Seeking diverse experiences and people, along with getting exposure to new groups and ideas, not only helps us to explore long-term career opportunities, but it also helps promote innovation and creativity, as discussed in Chapter 3.

9.2.4 Academic and Industry Orientation

Sometimes statisticians are categorized as academic or industry. This categorization reveals where a statistician works but may not reveal much about the nature of their work. Statisticians in academic settings do not just teach and do research, many focus on application. Similarly, statisticians in the industry may do a lot of research and/or do a lot of teaching and training.

9.2.5 Managerial and Technical Expertise

Especially in industry settings, statisticians may have the opportunity to focus on technical expertise (using statistics to solve problems and get work done) or managerial expertise (managing/supervising others who are getting work done). Typically, statisticians begin with technical roles, with some transitioning to managerial tracts after gaining initial experience. Those considering a managerial track should focus more on developing people skills than those who focus on the technical track. Nevertheless, people (and relationship) skills are important for all statisticians.

9.2.6 Size of Organization

The points made in the next section about utilizing coaches and mentors, and coaching and mentoring, lead to important considerations about the size of the company or organization where you work. In a small company, there is less opportunity for coaching and mentoring, but the coaching and mentoring could be more personal because it is less likely for an employee 'to get lost in the shuffle.' Regardless of the size of the company, understanding what opportunities for coaching and mentoring exist may be a good indicator of how well you fit. Organizations that make coaching and mentoring a priority often have more satisfied employees who make greater progress in their careers.

Another factor influenced by the size of the organization in which you work is the level of bureaucracy and time spent on non-core tasks. In a smaller company, you will have fewer non-core tasks. For example, in smaller organizations, you will have fewer emails and meetings, which can increase your focus and productivity. You may also have less required training, along with fewer regulations and SOPs in a smaller company. These factors can increase freedom and flexibility, thereby fostering greater focus and creativity. However, less guidance and regulation also increase the likelihood of making crucial mistakes or wandering off into unproductive approaches.

Yet another aspect related to the size of a company is the pressure for individual projects to succeed. Of course, all organizations seek success. However, small companies may have only one or a few projects and the failure of any one project can be catastrophic. On the other hand, big companies have more projects, and the future of the company does not rest on any one project. In fact, successful big companies often focus on failing fast so that time and resources can be devoted to other projects with greater prospects for success.

However, motivation and other cognitive biases discussed in Chapter 5 create pressure for the success of individual projects even in big companies. The pressure for success is not just an industry phenomenon. Pressure for success can be great in academic careers where a researcher may have one or two lines of research that must be 'successful' for continued funding and/or tenure.

As the old saying goes, there is no one size that fits all. However, the above discussion suggests that the newer the employee, the greater the likelihood that a better fit will be found in larger organizations that can provide greater structure and guidance, along with a wider array of opportunities, all of which can be advantageous early in a career.

9.3 COACHES, MENTORS, COACHING, AND MENTORING

Coaching and mentoring are terms that are often used interchangeably, but they are not the same. The role coaches play in the workplace is different from a coach on a sports team, although coaching can involve different tasks in different workplaces. In some situations, the role of a coach is to help a newer employee understand and develop competence in the fundamental aspects of their job; that is, to help train the new employee. However, coaching can also entail helping employees who have been on the job for a while with a specific aspect or aspects of their work.

A coach partners with the employee to help the employee achieve their goal(s). In many instances where the goal goes beyond initial training, the coach may not give specific advice, but instead asks questions and makes observations that increase the employee's awareness and commitment, and helps the employee find the answer that is right for them. When coaching goes beyond initial training, the goal of the coach is more to help the employee understand the questions and considerations that lead to the desired outcome rather than to state what needs to be done.

Coaching relationships tend to be of shorter term, often one year or less. Mentoring tends to be a longer-term relationship geared toward more general professional development than coaching, although the lines between coaching and mentoring can blur. Although coaches need to be more senior in experience than the employee they coach, mentors tend to be even more senior. Important aspects of mentoring that are less central to coaching include being a role model, providing a longer-term relationship focused on more general overall development, whereas coaching often focuses on one or two specific areas. Mentoring can involve less frequent and less structured sessions than coaching.

It is not uncommon, and often beneficial, for a mentor to come from outside the mentee's function or organization, thereby providing a broader perspective on issues of importance.

Both coaching and mentoring require trust, respect, confidentiality, and commitment by both parties involved. Casual chats at lunch every once-in-a-while may be useful breaks from work, but they do not replace coaching and mentoring. The coach or mentor's job is not to 'fix' anything or anyone, or to insist on a particular path forward. It is about helping the employee understand their choices and how those choices relate to their goals.

Some specific areas where a coach can be useful include:

- Address a specific behavior, habit, or lack of skill that is slowing an employee's development. For example, to help an employee speak up more in meetings or to improve presentation skills

- Support an employee with a stretch goal of taking on a new responsibility sooner than would otherwise be the case

- To supplement more formal training or development programs, for example, in leadership

- Inspire employees to maximize their talents

Some specific areas where mentors can be especially useful include:

- A role model for effective leadership or personnel management
- Increase cross-functional understanding and collaboration
- Increase diversity in thinking and perspectives
- Inspire employees to think about the long-term possibilities in their career and life

Having long, happy, and accomplished careers is complex and difficult. Utilizing coaches and mentors increases the chance of success.

Although the benefits of having a coach and a mentor are discussed frequently in self-help, personal development, and business books, the benefits of being a coach or mentor are less well known, but powerful. Being a coach or mentor is not only a great way to pay forward what others have given to you in the past, but it is also a great way to continue your own development. Some important aspects of mentoring are illustrated in the following quotes.

- 'A student was given a mentoring opportunity because having someone leaning on him would help him to stand even steadier.'   Thomas Hughes
- 'The mediocre (mentor) tells. The good (mentor) explains. The superior (mentor) demonstrates. The great (mentor) inspires.'   William Arthur Ward

A useful way to think of the benefits of being a coach or mentor is that while you may have acquired many skills and perspectives in your career, being a coach or mentor provides additional practice and diversity in applying those skills.

9.4 CONTINUED LEARNING

Continuing with the car metaphor for building a career that was introduced in Chapter 1, how do we maintain/improve the potential for raw power from our engine? If our prior experience does not account for everything we need to know, how do we continue to learn and grow on the job? When we are faced with the inevitable pressures of meeting timelines, where will we find the time for continued learning?

We may attend a conference every now and again and/or internal seminars and lunch and learns. But that alone is unlikely to sustain development. We can also learn how to learn as a routine part of everyday work. Methods to increase productivity were discussed in Chapter 2. Becoming more efficient in routine tasks saves time for learning. Increasing efficiency by 10% is not an outlandish goal but can have a big impact on learning because 10% of a work year is 25 days, some or much of which can be used for learning and development.

Learning and development can also be enhanced by incorporating learning into assigned work. For example, when assigned to routine deliverables, take a bit more time to learn the nuances of that task. Take time to understand the analyses and the inferences in detail, along with the science or business rationale underlying the hypothesis being tested. Take a bit of extra time to assess and understand competing methodologies for a particular application, if relevant. If this work reveals interesting considerations, present this work to your peers in a seminar or shared learning forum wherein you can practice your presentation skills; or, if applicable submit the work to a peer-reviewed journal or congress presentation.

To focus our discussion on continued learning consider two types of cultures: First, the winning culture. Many companies talk of their winning culture. 'We will win in the marketplace and ...'; 'We will beat the competition ....' In a winning culture:

- Performance is the most valued measure
- Reward and recognition for the most talented, who have the greatest impact today
- Rivalry is encouraged by comparisons
- Mistakes are criticized and corrected
- Willingness to do whatever it takes

Although these attributes might be laudable in some scenarios, such as sports, their applicability to research and business settings may be problematic. A winning culture often teaches that mistakes are to be avoided and it builds fear of failure. Yet, in research, we know many projects will fail. We should be willing to fail if we are to make progress and find breakthroughs. The willingness to do whatever it takes can be ethically challenging in regulated industries and wherever integrity is important.

Next, consider a mastery culture where we reward and recognize effort, learning, improvement – doing what it takes to be your best. Focus is on:

- Professional and personal growth
- Winning is highly valued, but winning takes care of itself if workers are engaged, dedicated, and improving their skills
- Focus on the process needed for success and success will come. Focus on success and we may forget the process

The cornerstones of a mastery culture are teaching and learning, mentoring, coaching, and leadership. There is an important connection between teaching and learning exemplified in the following quotes:

- 'In learning you will teach, and in teaching you will learn.'
  Phil Collins

- 'No one learns as much about a subject as one who teaches it.' Peter Drucker

- 'Tell me and I forget, teach me and I may remember, involve me and I will learn.' Ben Franklin

Learning how to learn on the job, having learning objectives as part of each year's performance plan, and sharing what you learn with others are all important aspects of career development. Understanding whether a prospective employer or group where you are considering working has an emphasis on winning or an emphasis on mastery is an important consideration.

9.5 PUTTING THE PRINCIPLES INTO PRACTICE

Continued learning while on the job is essential to career development, and career development is an essential element of having a long, happy, and accomplished career in statistics. However, continued learning is challenging given the day-to-day demands of most statistical jobs. Therefore, unless we plan efficient learning opportunities, little learning will take place.

Yearly performance plans should include learning objectives along with a plan for how these objectives will be met. Although some learning opportunities may exist through attending conferences, seminars, lunch and learns, etc., it is also important to seek opportunities to learn as part of regularly assigned duties. Taking a bit of extra time to ask questions, do background reading, and dig deeper into the nuance of relevant analytic techniques is often an efficacy way to learn. These learnings can then be reinforced by teaching what is learned to others.

These on-the-job learning opportunities are beneficial to individuals and to their organizations. The extra diligence in delivering assigned work helps ensure that work is of high quality, which benefits the organization and looks good on the individual's performance record, while also advancing the technical skills of the individual worker. Sharing key learnings with the organization also advances the individual's communication skills and increases institutional knowledge, which is good for the organization. This is a clear win-win situation.

Although it will always be challenging to find or make time for continued learning, utilizing the techniques to enhance productivity outlined in Chapter 2 make continued learning and maintaining work–life balance realistic.

When considering career development, longer-term planning, such as a three- to five-year plan is needed so that we can direct learning toward future goals and roles. These longer-term plans should take into consideration factors such as breadth versus depth of knowledge, the benefits from coaches and mentors, and the benefits of being a coach or mentor.

It is also important to develop time-efficient ways to learn. For example, audiobooks or podcasts while commuting can be a useful and time-efficient way to learn. Listen to or read a variety of material. Read or listen to books and podcasts for recreation or to recharge and for career development. Include topic areas of interest to you in addition to areas where you want to strengthen or broaden your skills.

An important part of career development that is not usually part of yearly or longer-term plans is impression management. Thought should be given to this topic because when impression management is done poorly the employee can be seen in a negative light for trying to 'suck up' to important people in the organization, to curry favor beyond their actual merit. When done properly, impression management results in the employee getting the credit that is deserved rather than being overlooked. A coach or mentor can be a good sounding board in developing an appropriate approach to impression management.

9.6 REAL-WORLD EXAMPLE – MY JOURNEY

Following is a list of my various work and educational experiences followed by a more detailed account of my career, focusing on events and learnings that map to key points in the preceding chapters of this book.

- 1976–1978   University of Missouri program in production agriculture

- 1979–1985   Co-owner of family farm

- 1988   BS Animal Science, Colorado State University

- 1990   MS Animal Breeding and Genetics, Colorado State University

- 1993   PhD Animal Breeding and Genetics, Colorado State University

- 1994–1998   Research Scientist & Assistant professor, Colorado State University, Department of Statistics

- 1998–2018   Eli Lilly & Company

- 2018–2021   Biogen

- 2021–2022   Cortexyme

- 2022–   Pentara

Today, it seems ludicrous that my initial choice was to pursue a career in farming while being legally blind. But I grew up in a small farming community where my family had farmed since the 1830s. Farming is what I knew and I did not consider other options. It was hard work and after seven years of high interest rates and no profits, I needed a change.

In 1985 I returned to college, first attending the local community college for a semester taking the basic 100-level courses in math, writing, etc., that were not part of my earlier studies in production agriculture. I did not have a firm plan for what to study or what career to pursue, but the two factors that drove my decision-making were that I already knew a lot about agriculture so continuing in that area, at least initially, made sense; and I knew that geographic separation would help me to move on from farming.

Therefore, I decided to attend a university with good programs in agriculture that was in or close to the Rocky Mountains. I had fallen in love with the mountains through my experiences attending training camps and competing in track and field and road racing. Colorado State University in Fort Collins, CO, and Montana State University in Bozeman, MT, both fit my educational and geographical criteria. In a stroke of good fortune, at this same time, my older brother landed a PhD internship at Colorado State; so, I packed my bags and moved to Fort Collins.

After all the years of hard work on the farm, college seemed easy. The hard work and dedication I learned through farming and athletics translated well to academics. An important factor was that I did not take heavy course loads. A standard course schedule was 15–16 credit hours per semester and the minimum for being classified as a full-time student was 12. I typically took 13 or 14 credits, which gave me time to study each subject thoroughly while maintaining a great work–life balance, with a bit of time left over to tutor student-athletes in chemistry.

I strung together good grades in diverse science courses, including Biology, Genetics, Chemistry, Anatomy, Physiology, and Organic Chemistry. The lowest grade on my undergraduate transcript was a B- in Introductory Statistics – and my grade was an overestimate of my true understanding! In fact, I was so scared of statistics that it was a mental obstacle as I approached what I anticipated would be my favorite undergraduate class, Animal Breeding and Genetics, for which statistics was a prerequisite. My primary role on the family farm had been developing and overseeing the genetic improvement of our livestock herds. Hence, Animal Breeding and Genetics was a natural fit.

However, my weakness in statistics motivated me to consult with the professor to figure out a 'survival plan.' The professor,

Rick Bourdon, who subsequently oversaw my graduate studies and became a key mentor, assured me that the key elements of statistics would be reviewed in the course and that I would survive. The statistics review in Animal Breeding and Genetics was covered in a few lectures during the second week of the course. In those two lectures, I learned more about statistics than in the entire semester of Introductory Statistics.

Rick Bourdon taught statistics as a tool to address important questions in Genetics, which contrasted with the instructor in Introductory Statistics who taught the calculations, but not the concepts and applications. With Rick, the light bulb turned on and I understood the concepts behind the calculations. This experience taught me the value of understanding that different people learn in different ways, and most of all it taught me the value of working hard to develop clear explanations. Something that previously had me completely baffled, Rick could explain in a few sentences with an example.

I was so motivated by this experience that I decided to pursue an MS and PhD in Animal Breeding and Genetics, with Dr Bourdon as my major professor.

Graduate school at Colorado State went much as my undergraduate studies. Focused work, dedication, and exposure to great teachers and thinkers along with a proper work–life balance led to success. The great teachers and thinkers included professors and other graduate students. Our grad student cohort interacted much as the 'Coffee House philosophers' in a hub and spoke model described in Chapters 2 and 3. We had many great discussions in our grad student office and computer lab, which helped everyone to learn and develop new ideas, which we then went off and worked on individually.

An essential part of my graduate training included teaching graduate and undergraduate-level courses and advising other graduate students outside of my core discipline on the design and analysis of their research projects. These experiences provided key foundations in communication and exposed me to a wider array of analytic problems. Although my formal statistical training lacked some of the key ingredients common to PhDs in statistics, my summative educational experiences included strong foundations in science, data analysis, and communication skills, which proved to be an excellent background for a career as a biostatistician.

After finishing my PhD, I applied for several academic positions while working as a research scientist and instructor in the Statistics Department at Colorado State. For many years, the Animal Breeding and Genetics group had a strong collaborative relationship with the Department of Statistics, which was further enhanced when a husband-and-wife team came to Colorado State as visiting professors, one in Statistics and one in Animal Breeding and Genetics. It was through these connections that I landed the non-tenure track position in Statistics. Although I had several academic job offers, none were the right fit, and I did not get the job I coveted in Animal Breeding and Genetics at Colorado State. However, I was enjoying my work in Statistics, teaching undergraduate Biostatistics to non-stats majors, and providing statistical consulting on clinical research projects in the Department of Veterinary Medicine. My role eventually became a non-tenure track joint appointment in Statistics and Veterinary Clinical Sciences.

The work with the Vet Med group was rewarding. The researchers appreciated my science background because I could more easily understand their work than other statisticians they had worked with who did not have a science background. They valued getting their work done on time and having it summarized in clear, concise reports. The experience I gained assembling databases, analyzing data, and writing reports on over 60 projects in a four-year period was invaluable for my later work in the pharmaceutical industry. Unfortunately – well fortunately, when a new department head took over in the Statistics Department, my non-tenure track role did not fit into his long-term plans. Hence, I needed to look for a job, and that is how I ended up at Eli Lilly without having trained for or seeking a career as a statistician in the pharmaceutical industry. How lucky can you be?

At Lilly, I joined a large team in which new members were coached and mentored, which helped me to apply the skills I had developed at Colorado State in a manner that was effective at Lilly.

Although my career had not gone according to plan prior to joining Lilly, at Lilly I developed mid-range and longer-term plans that followed a logical progression. I first developed a depth of expertise in the neuroscience therapeutic area, focusing first on later-phase clinical trials and then on earlier-phase trials. Gradually, I gained experience in other therapeutic areas through advisory/consulting roles and by being a member of several protocol review committees. After 10 years at Lilly, I had extensive experience in all four phases of clinical development and exposure to multiple therapeutic areas.

During those first 10 years, I developed my overall career goal. I wanted to be a good drug hunter who helped others to become good drug hunters. Although this was far from a specific, step-by-step plan, it provided guidance and a framework from which to develop more specific plans and to research and evaluate other opportunities.

From years 10 to 15 at Lilly, I was Statistics Group Leader in early-phase neuroscience, overseeing the statistical aspects of a portfolio that included about 20 compounds. This was an exciting time with new challenges. The optimum way to develop a group of drugs is different from the optimum approach to developing a single compound. These new challenges inspired new research projects and new collaborations, which, as in my earlier missing data and longitudinal analyses work, borrowed ideas from other fields.

After the neuroscience group leader role, I was ready for the role I had been working toward for 10 years – Technical Leader in the Strategy and Decision Science group. In this role, I led the summaries and evaluations of the company's entire portfolio and led research on ways to optimize the portfolio, with a focus on evaluating alternative prioritization schemes.

After completing my assignment in Decision Sciences, I was happy to return to the statistics function where for the next four years I worked in the immunology therapeutic area as statistics group leader. This was my first extensive exposure to immunology, and I enjoyed learning new things. It was a happy and successful period.

During this same period, I continued to consult broadly across all therapeutic areas in the company on a variety of statistical matters. I also continued my long-standing work in missing data, which included involvement in industry-wide working groups and continued to lead the part of the company's Innovative Analytics group that focused on missing data problems.

In the fall of 2017, the company offered a voluntary early retirement program. Initially, I thought that would be a good offer for some people. Then, I approached the issue from a different angle: 'Under what conditions would this be a good opportunity for me?' I did not want to stop working, but if I could find the right role, this might be an opportunity to try something new.

I had strong statistical connections at Biogen, including with the Vice President of Statistics who had been my first coach at Lilly. She and I worked out a role that in concept would be to help the statistical organization mature to scale up the successes they had as a small unit into a mid-sized unit. The company had recently gone through a period of rapid growth and success. My earlier work in portfolio management suggested that the approaches to drug development that lead to success in a small company were unlikely to be successful in a larger company and I wanted to help Biogen and its stats group mature into its newly acquired larger size.

Several initiatives that I was involved in helped to progress the company's approach to portfolio management. And I became involved in the development of a drug for Alzheimer's disease a day or two after development of that drug had been terminated by an interim futility analysis. The subsequent steps that led to the eventual approval of the drug, but its lack of commercial success, are controversial. I will not go into those details here other than to say the issues presented major statistical challenges that were difficult and exciting to work on. The drug's team leader and I developed an effective and collaborative working relationship and she also became a valuable mentor.

After the intense work and ultimately disappointing result on the Alzheimer's drug, for the first time in my career I was burned out. The work on the Alzheimer's drug had despite my best efforts thrown off my work–life balance. I reduced my workload to an 80% role as VP of statistics at a small biotech.

It was interesting to work at a small company, being much closer to the inner workings of upper management but also not getting as much interaction with other statisticians. The company had only two clinical drugs. When a major safety snag beset the lead compound, it was time to look for another job. I ended up working in the statistical consulting group at a CRO. Given that role is just starting I cannot say much about it. However, it is likely I will continue to have some successes and some failures – and it is likely I will continue to enjoy what I do and I hope to make meaningful contributions.

As I reflect on all these experiences, I am proud of the drug development work and statistical methodology work I have done. I am thankful for the opportunities and help I have had along the way. With the wisdom of hindsight, I am not much troubled by my setbacks (see real-world experiences in several chapters) because I did my best. At times, my best was not good enough, but that is neither surprising nor disappointing because I had the great fortune to work on important projects and drug development is hard. Failure is part of the game.

As presidential historian Doris Kearns-Goodwin noted (Kearns-Goodwin, 2018) in her book on Leadership that was discussed in Chapter 7, resilience is a key attribute for leadership and success. Kearns-Goodwin defines resilience as the ability to maintain motivation through disappointment. My metaphor for resilience is that it does not matter how many times you get knocked down; it matters how many times you get back up. And, due to good fortune and the great support from colleagues and family, I am still standing!

CHAPTER 10

Craig's List

ABSTRACT

This chapter is based on a paper the author wrote recently that contains 40 quotes. These quotes and the explanations associated with them provide memorable reinforcement of points made throughout earlier chapters in the book.

10.1 BACKGROUND

This chapter is based on a reflective period in my career. The paper titled Craig's List (Mallinckrodt, 2019) was the product of that reflection. The paper is a collection of sayings. As noted in the introduction to the paper, few of these sayings are my original thoughts, and I can only rarely cite the source for the others. The intent of reviewing these sayings here is that although simple one-liners cannot fully explain what to do and how to do it, they can be convenient reminders of the more detailed points made throughout this book.

10.2 THE LIST

The quotes and sayings from Craig's List (Mallinckrodt, 2019) are summarized below. A few of the quotes were used earlier in the book, but they are worth repeating here. Some of the sayings are

DOI: 10.1201/9781003334286-13

self-explanatory. For others, additional comments are provided, including how these quotes illustrate key points and map back to material covered in earlier chapters. The quotes are arranged into groups according to general topics.

10.2.1 Technical Acumen

Although this book does not deal with developing technical acumen, several sayings help focus on useful ideas.

> 1: The difficulties lie not so much in knowing the principles, but in putting them into useful practice.

Although this saying may seem to be directed at applied statisticians, it describes a fundamental challenge faced in many scientific disciplines. In fact, the saying is attributed to Robert Bakewell circa 1750 and suggests that the value in knowing any theory is greater if we know how to use that theory.

> 2: If a result does not look right, act like it is not.

It is often easier to develop a new theory or in some way explain an unexpected result than it is to believe a mistake was made. However, the possibility of a mistake cannot be ruled out. Therefore, when we have an unusual or unexpected result, it may be time to double down on validating the correctness of that result.

> 3: Things only become obvious after they become obvious.

In retrospect, the solution to a tough problem, the bug in code, etc., may seem obvious, but before we find them, they are not easy to spot. This is an example of the hindsight fallacy (Chapter 5) in action.

> 4: The quest for perfection gets in the way of good enough.

Time devoted to a task often has diminishing returns. Therefore, at some point, our time would be better spent elsewhere.

Understanding when to move on requires an understanding of priorities (Chapter 2).

> 5: I do not care what you do or how you do it, so long as it has the highest probability of giving me the right answer.

A Vice President of early phase research told this to me at a time when there was debate about appropriate ways to handle missing data in clinical trials. I advocated for a 'newer' approach known as MMRM (mixed-effects model for repeated measures). However, regulatory precedent was still anchored in LOCF (last observation carried forward). I appreciated the VPs willingness to delegate the decision along with his being clear as to the criterion upon which that decision should be made. More generally, this quote reminds us that we should not choose methods to advance the cause of those methods, or to mindlessly adhere to tradition. Our choice should be based on what works best.

> 6: Do not talk to me about averages – because if you put your head in the freezer and your butt in the oven, on average you should be comfortable.

This quote came from my father. I used it many times to explain to diverse audiences the importance of understanding variation in outcomes. The quote was useful because it put a complex idea into simple terms that everyone understood. Even 20 years after hearing this explanation people remember it. This is an illustration of how stories can be useful in conveying complex ideas (Chapters 4 and 5).

10.2.2 Philosophical Approach to Work and Career

> 7: Keep your eye on the football.

The 'football' is the shape made by the overlap of circles in a Venn diagram. The point is to figure out what is important to your team or organization and figure out what you like to do. Then spend as much time as possible where these two spheres overlap. This is another example of prioritization (Chapter 2) as well as an example of career planning and development (Chapter 9).

8: Be a multiplier.

If you work as effectively as you can, you contribute 1X of production. If you can also help 10 others to work 10% more effectively your contribution is 2X of production. We do not have to do everything ourselves. The ability and willingness to work with and through others can allow us to accomplish more than we could do on our own. If you believe that the only way to do things right is to do them yourself, you are destined to not get much done. This idea is an extension of ideas on influence and leadership (Chapter 7).

9: Take what you find and make it better.

10: Some look at the way things are and ask why. He looked at the way things ought to be and asked why not.

Saying 9 is from a founding family member of Eli Lilly & CO, my long-time employer. Ten is from Senator Ted Kennedy's eulogy of his brother Robert. These sayings remind us that it is easy to complain but complaining does not help get things accomplished. We all get frustrated and need to vent. However, we should limit the venting to a short amount of time and then turn our attention to solving problems and making things better. This point relates to ideas on team dynamics discussed in Chapter 6.

11: You cannot eat the apple in one bite.

12: The man who moves a mountain begins by carrying away small stones – Confucius.

Sayings 11 and 12 advise us to break down big tasks into component pieces so that the overall task is not too intimidating to begin. Doing the research and writing of a manuscript for a peer-reviewed journal, writing a book, conducting and/or analyzing data from a clinical trial, or many projects in business, are all intimidatingly larger tasks. But if we take it one step at a time, with a well-thought-out plan, we can get the job done.

> 13: The best time to plant a tree was 30 years ago. The next best time is today.

There are many things I wish I had done earlier in life. But I am still very glad I did them later in life. Not getting an early start is no excuse for never doing it.

10.2.3 General Advice

> 14: Do not compare salaries or promotions. These comparisons lead to greed and jealousy, and those will not help you get a raise or a promotion.

> 15: Success requires doing things right and doing the right things.

> 16: Greatness is not one thing, it is many little things all done well.

> 17: Practice makes perfect permanent.

The old saying is that practice makes perfect, but practice does not lead to a high-quality performance or result unless the practice is of high quality. Practice engrains actions, responses, habits, etc., whether they are perfect, good, or bad. Therefore, to improve we should practice with diligence and purpose.

10.2.4 Perspective

> 18: Good judgment comes from experience, and experience comes from bad judgment.

19: Quitting is a permanent solution to a problem that may be temporary.

20: Attitude and approach are more important than aptitude.

21: When one door closes, you need to open others.

22: I did not get the things I wanted. I got the things I needed instead.

These sayings remind us that our careers have ups and downs. We make mistakes. We have failures. We get discouraged. How we respond in these downtimes says more about our character and our prospects for success and happiness than how we respond in good times. These sayings are another way of highlighting the importance of resilience that was covered in Chapter 7.

10.2.5 Performance and Productivity

23: Many people think that giving a good presentation is glamorous, but nobody thinks the preparations required to give a good presentation is glamourous. (Ditto for writing good papers.). This saying reinforces the points on communication made in Chapter 4.

24: When I started, I thought time management was the most important skill. Now it is attention management.

25: Make time to think, every day. Take a walk outside, every day.

26: I can do 12 months of work in 11 months, but I cannot do 12 months of work in 12 months.

These quotes are another way of looking at the ideas covered in Chapters 1, 2, and 3 on work–life balance, productivity, prioritization, creativity, and innovation. Today's work and social environment pose challenges to our ability to sustain long, successful, and happy careers. One of the most useful things I did was to study how to work efficiently and creatively in today's distractible work environment. I encourage you to do the same.

27: Find the 10%.

Most (all?) of us feel the pressure to deliver the routine parts of our jobs and have little time set aside for career development, continued learning, or researching topics in which we are interested. A 10% increase in the efficiency of our work may not seem like much, but it yields the equivalent of 25 extra days each year to do other things. This idea is central to points covered in Chapter 2 on productivity, Chapter 3 on innovation and creativity, and Chapter 7 on continued learning.

10.2.6 Learning

28: You have the perfect background for this position. The other 90% of what you need to know you can learn on the job.

This quote was used in the preface to this book. As noted in Chapter 9, sustaining productive and happy careers requires continued learning on the job, which is challenging. However, learning new things also presents opportunity.

29: In learning you will teach, and in teaching you will learn.

30: No one learns as much about a subject as one who teaches it.

31: Tell me and I forget, teach me and I may remember, involve me and I will learn.

Quotes 29–31 point to active learning and teaching for individual and organizational learning. The goal is to identify an important problem, study it, or conduct research to solve the problem, and then teach what you learned to others. It is a win-win for you and your organization (Chapter 7).

32: A student was given a mentoring opportunity because having someone leaning on her would help her to stand even steadier.

I cannot say how much those I coached and mentored gained from the experience; I know that I gained a lot. See Chapter 9 for more on coaches, mentors, coaching, and mentoring.

10.2.7 Leadership

33: The mediocre leader tells. The good leader explains. The superior leader demonstrates. The great leader inspires.

34: Your most important task as a leader is to teach people how to think and ask the right questions so that the world does not go to hell if you take a day off.

35: Leaders should influence others in such a way that it builds people up, encourages, and edifies them so they can duplicate this attitude in others.

Many books on leadership have been written. These sayings summarize key themes that were covered in more detail in Chapter 7. Leadership can be intimidating because most of us do not feel that we have the inherent skills to be a leader. However, as noted in Chapter 7, leadership skills can and must be learned, even for the most gifted. It is reassuring to know that if you care about the people you lead, you will be a good leader regardless of your leadership style.

10.2.8 Ethics and Behavior

36: The night before your drug or product hits the market, it is comforting to know that you did your best to find the right answers.

37: When you are trying to figure out what to do, ask yourself, how would your family feel about your actions if they read about it on the front page of *USA Today*.

Perhaps the quote should be updated since so few people read newspapers today. So rather than 'read about it on the front page of *USA Today*,' insert 'went viral on social media.'

Ethics is a complex topic that was not covered in detail in this book. However, if you have done your best, and those whose opinions you value are proud of your actions, you must be on the right track.

38: On your last day at work, your colleagues will not recall how many papers you authored or how many submissions you did. They will remember how you made them feel.

We cannot go through our careers with the only goal of having people like us, but if people like us are more likely to achieve our career goals – and be a lot happier. This idea relates to the principle of liking (and other principles) on influence outlined in Chapter 7 and to principles of working relationships covered in Chapter 8.

10.2.9 Priorities and Work–Life Balance

39: If work is the most important thing in your life, you should work on your life.

40: The days are long, but the years are short.

Quotes 39 and 40 relate to work–life balance discussed throughout the book. For many, work will be more rewarding if it is part of a balanced life, and it will lead to more rewards outside of work. Use your time wisely because it will pass quickly.

41: Answer these three questions: Am I motivated in my work; am I working smart; am I working well with others?

If all three answers are yes, you are on track for success. If one or more answers is no, you should probably make changes. If you do not know the answer to one or more, you should seek feedback. Be careful, though. In today's work environment, working too hard is common. It may be useful to substitute 'Am I motivated in my work' for 'working hard' because motivation ensures we are working hard enough.

42: Be sure to enjoy this day.

This advice came from my wife Donna on the morning of the most meaningful presentation of my career. The preparations were exhausting, and I had lost perspective. Donna's advice helped me to appreciate how lucky I was to be part of an important moment. This perspective allowed me to overlook the fatigue and burden, and therefore appreciate the opportunity I had to help make a difference. Emotion and attitude are powerful forces. Learning how to focus them in positive directions leads to greater success and satisfaction.

43: Success is a journey, not a destination

10.2.10 A Final Word, or Two

'It's a good question to be asking yourself

What is the good life, what is wealth?

What is the future I'm trying to see?

What does that future need from me?'   *Jackson Browne*

'Live the day. Do what you can.'   *Wesley Schultz*

Road Map

ABSTRACT

This chapter summarizes key ideas from each of the preceding chapters. The intent is more than to reiterate key points. By presenting key points in a more condensed format, it is easier to see how the pieces fit together. Continuing with the careers as a car analogy introduced in Chapter 1, a career, like a car, must be more than a collection of components. Regardless of the quality of each component, the overall performance of a car is determined by how well the components work together, and so it also is with careers. We need good individual components, but to have a long, happy, and accomplished career those components must function well together. Chapter 11 focuses on those connections so that we can fit the components of our careers together in the most effective manner.

11.1 INTRODUCTION

This chapter summarizes key ideas from each of the preceding chapters. The intent is more than to reiterate key points. By presenting key points in a condensed format, it is easier to see how the pieces fit together. Continuing with the careers as a car

DOI: 10.1201/9781003334286-14

analogy introduced in Chapter 1, a career, like a car, must be more than a collection of components. Regardless of the quality of each component, the overall performance of a car is determined by how well the components work together, and so it also is with careers. We need good individual components, but to have a long, happy, and accomplished career those components must function well together.

Technical acumen is the engine of our careers but being good at statistics is not enough in most scenarios. As we say, it is necessary but not sufficient. In fact, some of the most unhappy statisticians I have known are those with strong technical ability but who felt they were unproductive or unable to influence. Besides technical acumen, statisticians benefit from the following skills:

- Work productively, prioritize, and maintain work–life balance

- Creativity and innovation

- Communication, both oral and written

- Think critically and to make decisions under uncertainty

- Influence and leadership

- Relationship skills

- Career development and continued learning

We do not have to be experts in all these aspects. When we pick a car, the decision is based on what features we value most. In careers, it is up to us to decide what features or combinations of features are most important at a particular point in our career. The features we wish or need to emphasize often change over time. Productivity and prioritization are foundational to all skills because without them we will not have time to develop the other skills and maintain work–life balance. Therefore, the best time to develop skills in productivity and prioritization is early in our career, the second-best time is now.

11.2 PRODUCTIVITY, PRIORITIZATION, AND WORK–LIFE BALANCE

The Workism culture so prevalent today, with its 24/7 connectivity results in us working too long and too hard, which in turn leads to focusing on activity rather than accomplishment. If we are motivated, we will be mindful of work goals and we will work hard enough. Although some statisticians would benefit from working harder, for most of us working harder is not the answer. Working smarter – with better work–life balance – is the answer.

However, if you are not motivated and engaged in your work you are not on track for a long, happy, and productive career. You probably won't have the energy and discipline needed to work smart and to work well with others. If you are not interested in and inspired by the work in your current job, or if the main demands of the job do not match your interests or skills, you may fare better in a different role. But be careful. Quitting is a permanent solution to a problem that may be temporary. Unnecessarily changing jobs or roles may result in losing out on key learnings and insights because a project was not followed through to completion. On the other hand, if you are not generating any momentum, then it may be time to consider alternatives.

Long-term happiness is unlikely to come from acquiring things we want. Psychological research shows that once we get what we want, happiness sooner or later, and more likely sooner, wanes because we become accustomed to what we acquire. Therefore, we should have bigger-picture goals, an overall purpose in work and in life.

Successful people tend to be clear about their role in the events of their life. Successful people realize no one succeeds on their own. Being successful does not mean we were smart enough and good enough to do it on our own. Being successful requires being smart enough and good enough to seek out and work with others who make us and our work better – and being willing to help others become better.

Attention management is a foundational skill for statisticians, and for many others, because it is the base upon which other skills are built. With good attention management, we can:

- Implement the Deep Work/Hyperfocus skills central to efficient productivity. Being efficient and productive frees up time for:
- Scatterfocus time for creativity and innovation
- Developing effective presentations, manuscripts, and other communications
- Critical thinking, which is an essential part of making decisions under uncertainty
- Developing effective influence and leadership approaches
- Investing in work relationships
- Continued learning

Chapter 2 outlined the following four general approaches for the Deep Work/Hyperfocus needed for productivity.

- Monastic approach: Radically minimize shallow work in your life
- Bimodal approach: Schedule long stretches of time for isolation and Deep Work (days or weeks)
- Rhythmic approach: Schedule time every day to do Deep Work – i.e., manage attention/distractions
- Journalistic approach: Whenever you can switch into Deep Work (very hard to switch like this)

In many work settings, the rhythmic Deep Work strategy is most applicable. The key tactics outlined in Chapter 2 for rhythmic Deep Work were:

- Limiting social media and infotainment

- Limiting availability on IM/texts

- Turning off email notification and batch process emails

- Limiting multitasking

- Use quiet places away from your workstation for focus time

Because distractions are inevitable no matter where we work or what Deep Work strategy we employ, managing distractions is central to managing attention. Chapter 2 provided the following guidance on managing distractions.

- For fun distractions that can be controlled, set limits in advance such as checking Facebook for x minutes during lunch.

- For distractions that can be controlled and are not fun, plan means to avoid or minimize distraction. For example, when working on a manuscript plan to work from home to avoid distracting conversations with coworkers in an open office setting.

- For fun distractions that cannot be controlled, enjoy the moment but be ready to refocus.

- For distractions that cannot be controlled and are not fun, develop a coping mechanism such as wearing headphones to block noise and to signal to others you are focusing.

Deep work/Hyperfocus is hard, and therefore should be scheduled for when you have the most energy. For morning people, likely morning is best. For night people, nighttime is often best. During extended blocks of Deep Work, take short breaks to recharge.

Deep Work does not mean working alone. Consider a real or hypothetical hub and spoke layout. In hubs, interact with people and gain exposure to ideas and approaches. Then use Deep Work approaches in locations where distractions are minimized (spoke) to work on what was encountered in the hub.

To manage time and attention to generate more Deep Work, prioritization is key because it provides the opportunity for focus. Two important impediments to prioritization and focus are multitasking and difficulty in saying no to requests. Chapter 2 discussed why multitasking reduces productivity even if it feels like we are getting more done. Saying no is hard because we do not want to disappoint our boss or coworkers. However, when we say yes to something, we cannot manufacture the time to do it. Therefore, something else must be left undone. Hence, it is imperative to understand that we must say no sometimes to focus on the most important things, which will lead to the most total impact in the long run.

11.3 CREATIVITY AND INNOVATION

Creativity and innovation are not static attributes. They can be increased by understanding where good ideas come from and how the brain works in processing those ideas. The conceptual model for increasing creativity and innovation is to collect many dots and then spend additional time connecting those dots.

We should not wait for good ideas to come to us, we should go to where good ideas might be.

We can collect more dots through exposure to networks, such as groups, blogs, seminars. Anywhere we may be exposed to new ideas can be a network. Some worthwhile connections to ideas or information are serendipitous, but this is not the same as pure luck. By exposing ourselves to places where connections are likely, we increase the probability of serendipitous interactions.

We might find good ideas in regular team and functional interactions, or by joining industry work groups, journal clubs, etc. Making time for hallway conversations at conferences and at work can lead to new connections and ideas. Of course, the internet is a great way to gain exposure to ideas and to see connections between ideas. In addition, employees who work across areas solve more problems than those who focus on a narrow area. Therefore, problem-solving capability can be enhanced by seeking diverse work opportunities.

Reading is another great way to form new connections. Deep dive reading is especially useful because we read a volume of material in short order and thus have the ideas in mind more than from readings long ago. However, for many, schedules do not allow deep dive reading. Therefore, as we read it is useful to store key ideas from each reading in a 'common book' and periodically revisit the common book.

The hub and spoke model is a good way to collect and make use of ideas. We gain exposure to ideas in the hub and then go to the spoke to do individual work, including work that follows up on the ideas encountered in the hub. Hubs can be common areas at work, seminars, lunch and learns, conferences, the coffee shop, etc.

Controlling attention on one thing is Deep Work/Hyperfocus. The structured mind wandering of Scatterfocus is relevant to creativity and innovation because this is when our brains can make new connections, when we connect the dots we have collected. Walking, especially in nature, showering, work breaks, vacations, or working from an unfamiliar location, all provide breaks from the normal routines of life where the phase-lock part of the brain dominates. New and unfamiliar settings result in less time in phase lock, allowing for the formation of more new connections and the cultivation of ideas in the subconscious mind.

Good ideas may seem like, but seldom are, eureka moments. Instead, good ideas usually evolve over time as slow hunches. Therefore, creativity and innovation can be enhanced by interacting with others doing similar things, and by allowing time to let the ideas develop, including through refinements generated via interactions with others. For example, say you have an idea for a simulation study to investigate a new method or a novel implementation of an existing method. It can be useful to conduct a 'pilot' set of simulations, knowing the results of these simulations will not be part of the final product but will be used to inform how the final simulations will be done. Doing the pilot simulations gives our minds time to further evolve slow hunches, to refine our ideas.

Another approach is to write down all your ideas, along with key ideas of others, interesting quotes, observations, etc., in an indexed format as many great scientists have done over the centuries. In writing things down, especially when near other good ideas, hunches have a chance to become fully formed ideas.

Slow multiprocessing, which is different from multitasking, is the sequential processing of concurrent projects. It is yet another way to form connections as the mind moves across multiple problems to make connections. However, having a useful idea is only the start. We usually need to convince others of the merits of this idea before the idea leads to tangible benefit. Convincing others requires the ability to influence (covered in Chapter 7) and to communicate the idea (covered in Chapter 4). This is an example of how the components of a successful career must interact with each other, much as the components of a car must interact with each other.

11.4 COMMUNICATION

Chapter 4 provided detailed tips on creating and delivering presentations and on effective writing. These tips will not be reviewed here, except to highlight the most important aspects. You do not have to be Abraham Lincoln or Martin Luther King Jr. to get your points across, but you do have to be clear. If you have something important to say and you say it clearly, it will not matter that much if it is read verbatim off a cue card or delivered with soaring rhetoric.

The tendency is to tell others everything we know about a topic, or everything we did in researching a problem, in the order we did it. The ordering of events and the depth of knowledge are important to us. However, clarity requires that key points be explained in a manner understandable to the audience. And, the audience should appreciate why those points are important to them.

Many of the tips in Chapter 4 came from a guide to TED talks, which can be thought of as major presentations, such as an invited talk at a conference. However, presentations that are part of our routine responsibilities add up to have a major impact over time. These routine presentations afford opportunity to practice the principles noted in Chapter 4.

The same ideas apply to written communication. Many of the tips in Chapter 4 are geared toward scientific manuscripts or internal, technical documents. However, the principles can be applied and skills refined via day-to-day communications, such as emails or project update reports.

In developing the first draft of a presentation or a paper, carelessness will result in a greater need for revision. However, too much effort spent on refining content at the initial stage may inhibit assembling the most useful content. No matter how careful we are with an initial draft, much work will remain to be done. Hence, do not try to make the first draft perfect. Focus on getting the main ideas down, knowing revisions will be necessary.

Feedback from others is especially useful in assessing whether the original content assembled for a talk or paper is appropriate for the circumstance. Feedback is especially beneficial in making sure explanations are clear.

Read and study *The Elements of Style*. It is a concise, practical guide to good writing.

11.5 CRITICAL THINKING AND DECISIONS UNDER UNCERTAINTY

Critical thinking and making decisions under uncertainty is where the transition begins from working with self (topics covered in Part 1) to working with others (topics covered in Part 2).

Practical applications of critical thinking and decisions under uncertainty begin with understanding that individual, group, and organizational level factors each come into play. Regarding individual-level factors, humans are not wired for making good decisions under uncertainty. Cognitive biases are the primary source of poor decisions and even the most intelligent scientists are prone to these biases.

Chapter 5 covered key cognitive biases that can lead to poor thinking and poor decisions. Being aware of cognitive biases and how they influence individual thinking can help but will not eliminate the biases. The source of many of the biases can be found in 'System 1' thinking, where we use heuristics (rules of thumb) for making quick decisions. In 'System 2' thinking, we slow down, contemplate, evaluate, and calculate.

In critical thinking and making good decisions, it is useful to understand when each system should be used and to control our attention accordingly. We cannot contemplate, evaluate, and calculate every decision we make. Some decisions require instantaneous action. If we are crossing the street on foot and we see a car approaching as if they will run the red light, we don't have time to estimate the car's speed and its distance from us to know whether we should begin running. The caveman did not have time to consider whether that noise in the bush was a small animal or a large predator stalking him for lunch. Moreover, we make too many decisions to consider them all carefully. How much time do you want to spend on cost–benefit analysis and a quality decision-making process to evaluate every option for lunch?

Hence, the key is to recognize when we are in a situation where System 1 thinking can lead us astray. In those situations, we benefit from the ideas detailed in Chapter 5 to prevent biases from creeping into our individual thinking. To have the time available for such contemplation, we need to be productive, and to make the most of that time we need structured thinking – both of which require focused thinking from Deep Work/Hyperfocus attention management (Chapter 2). The solution to a problem may require a creative new idea or a novel approach, which will require collecting and connecting dots (Chapter 3). This is another example of how the individual career components work together.

Chapter 6 provided ideas on how groups can interact and think collectively better than any individual could on their own. In group settings, it is important to draw out the thoughts of everyone involved. Leaders and senior team members should be mindful of holding back their opinions and thoughts until all relevant information has come to light because if they express an opinion too early it will tend to stifle further debate.

In group and individual settings, it can be useful to adopt multiple frames when examining a situation. By defining problems in several ways, we can understand how each definition (frame) tilts us toward the same or different solutions. We can be more confident in a solution if we arrive at the same solution across multiple frames. If the best solution is frame-dependent, extra effort may be needed to ensure the most applicable frame is used.

Specifying assumptions, including implicit assumptions, and testing those assumptions and their impact on inferences are other keys to understanding the robust versus context specificity of a solution.

It is also important in both individual and group settings to distinguish between confidence and accuracy. Confidence and accuracy are not the same things. In fact, research has shown that greater confidence is associated with poorer decisions because confident people or groups may have overlooked crucial aspects of the situation.

Conclusions drawn from intuition (System 1) fuel overconfidence. Just because something 'feels right' (intuitive) does not make it right. System 2 is needed to slow down and examine our intuition, estimate baselines, consider cognitive biases, and evaluate the quality of evidence. It is important to understand in which situations intuition is useful versus likely to be misleading. In many scientific and business applications intuition can be misleading.

In group settings conflict is not always bad. In fact, it is often needed. Therefore, it is useful to understand the difference between the two forms of conflicts. Cognitive conflict is task-oriented disagreement or debate about issues and ideas. Affective conflict is emotional and personal. It is about personality clashes, anger, and personal friction. Hence, groups benefit from cognitive conflict and are hindered by affective conflict.

When a group is confronting an important decision, the best first step is often deciding how to decide rather than immediately trying to solve the problem or make the decision. Chapter 5 outlined processes that help avoid group think, stimulate cognitive conflict, minimize affective conflict, and avoid premature convergence on a single idea.

Although most individuals will not have the ability to change the culture of a group, much less an entire organization, it is important to understand whether the group or organization has the culture of yes, the culture of no, or the culture of maybe. Knowing if these conditions exist can help navigate attempts to get new ideas or approaches accepted, even if these bad attitudes cannot be changed.

11.6 INFLUENCE AND LEADERSHIP

Influence is an inherent part of leadership. Chapter 7 details how influence can be applied for good or bad. Of course, our focus is on the good, but being aware of how bad Agents can influence us is useful. Moreover, influence is not just for leaders. Every statistician or scientist who has made a key inference has a key idea, or advocates for a novel approach need to convince others on the merits of the idea. The quality of the idea will depend in part on using the attention management skills discussed in Chapter 2 and on our ability to collect and connect the dots discussed in

Chapter 3. However, the idea will not sell itself, we will have to convince others.

Our ability to influence can be increased by understanding and implementing the influence tactics and understanding the sources of power discussed in Chapter 7 and by clear communication (Chapter 4).

A part of good leadership stems from the ideas on critical thinking and decisions under uncertainty discussed in Chapters 5 and 6. Leaders and senior team members should promote an organizational environment that allows teams to function well. This includes avoiding the cultures of yes, no, and maybe. It also entails avoiding the five dysfunctions of teams (see Chapter 6) and promoting an environment for individuals to do Deep Work (see Chapter 2), to be creative (see Chapter 3), and to think critically and make decisions under uncertainty by putting systems in place that avoid cognitive biases (see Chapter 5).

Aspiring leaders should take advantage of the many team and functional meetings that provide opportunities to practice the influence skills outlined in Chapter 7. By consistently focusing on these opportunities, which are most often of smaller scale, and soliciting feedback, influence and leadership skills can be cultivated, thereby preparing aspiring leaders for large-scale opportunities for influence and leadership later.

In situations where intuitive thinking is useful, leaders should explain their intuition and the rationale behind their decisions for employees to understand and trust the decisions. Decisions based on intuition require more effort to build the trust in employees needed for successful implementation because employees will not share the leader's intuition.

Other things leaders should focus on include:

- Provide a process road map at the outset of the decision process

- Reinforce an open mind-set

- Engage in active listening

- Explain their decision rationale

- Explain how others' inputs were employed

- Express appreciation for everyone's input

The prospect of leading a group, even a small group, can be intimidating. Many potential leaders worry that they lack the inherent skill required to lead. Although some have greater inherent leadership abilities than others, much of what good leaders do can be learned. Most importantly, if you care about the people you lead, you can be a good leader using any number of leadership styles and a variety of skill sets. If you do not care about the people you lead, you are unlikely to be a good leader no matter what style you use or how smart you are.

Aspiring leaders can improve skills and increase confidence through leadership opportunities outside the workplace and within the workplace in ancillary roles. Consider, for example, a coaching role in youth sports, leading a volunteer organization, or organizing a seminar series at work.

11.7 EFFECTIVE WORK RELATIONSHIPS

The benefits of getting along with coworkers and the consequences of not getting along are substantial. Good working relationships help us and our teams to be more effective and promote greater satisfaction and happiness. Not getting along does the opposite.

Good relationships do not just happen, we need to work to promote them. Loneliness is common in the workplace, and the increase in remote and gig work exacerbates this problem by reducing in-person interactions. Burnout is on the rise (Fisher & Phillips, 2021). Moreover, being digitally connected 24/7 and deluged with information and communication, paradoxically, increases alienation. The 'Workism' lifestyle of always being online and always having at least part of our mental energy devoted to work isn't sustainable long-term. Chronic overwork eventually decreases output and causes stress-related problems such as burnout, fatigue, anxiety, depression, heart disease, sleep disorders, and strained relationships at home (Fisher & Phillips, 2021).

Chapter 8 provided the following tips on avoiding the destructive Workism mind-set:

- Well-being is more than physical health. It is mental and physical welfare, including feeling secure, accomplished, and personally satisfied.

- Money does not define success.

- Rest is not a sign of weakness. Taking time to recharge boosts productivity and resilience and guards against burnout.

- Burnout does not go hand-in-hand with success. The human cost of working long hours and high stress is severe. Success at work does not require sacrificing your personal life.

- Loving your job shouldn't mean sacrificing everything else in your life.

- The Workism culture is powerful and we have to actively work against falling into it.

- Create a personalized well-being plan to manage career pressures.

Strong social bonds are important in combating these workplace trends. When people feel comfortable, connected, and valued at work, business prospers. Connection, comfort, and contribution promote feelings of belonging and result from strong interpersonal relationships. Unfortunately, many team leaders and companies value these qualities less than productivity and performance, which flaw that is exacerbated by erroneous ideas on what is required for productivity. When well-being and healthy relationships are secondary to productivity, it is like a sports team of talented players who do not stay healthy and do not work together. Companies and teams that prioritize well-being and healthy relationships create the strongest, healthiest team dynamics.

Chapter 8 provided tips on establishing effective working relationships, such as

- Initiating conversations by asking questions
- Saying thank you
- Being positive
- Introducing yourself and others
- Letting others know who you are
- Sharing information
- Supporting the work of others
- Asking others to get involved in your work
- Initiating repeated interactions
- Participating with others in non-work activities

Giving and receiving feedback is difficult but important. We will improve more, help others improve more, and get along better with our coworkers if we can effectively make use of and give constructive feedback and advice. An overarching principle in giving and receiving feedback is to anticipate – and avoid – strong emotional responses. It is best to hold off on giving feedback or responding to it until it can be done in a calm and rational manner.

Chapter 8 provided the following tips on giving and receiving feedback.

- Be professional, not personal

- It's a two-way conversation

- Focus on facts rather than feelings

- Be direct

- Balance positives and negatives

- Choose words carefully

- Focus on fixing not finding fault

Impression management can seem like 'brown-nosing' or 'sucking up.' When done properly, however, impression management is helping others to see who you really are. Common methods of impression management include the choice of clothing, the avatars or photos used to represent ourselves online, descriptions in résumés and online profiles, and how we comport ourselves in the workplace. Appropriate impression management builds credibility, maintains authenticity, is believed by others, and fosters a variety of favorable outcomes for the individual, his/her team, and the organization.

Not all aspects of our true self must be disclosed in the workplace. However, trying to win social approval by suppressing too much of one's true self can lead to psychological distress and unfavorable outcomes. It is important to keep in mind that whether we manage our professional image or not, coworkers are forming impressions. They watch our behavior and draw conclusions about the kind of person we are, whether we are trustworthy or whether we are dedicated to team goals, and how we will react in difficult situations.

11.8 CONTINUED LEARNING

Continued learning while on the job is essential to career development, and career development is an essential element of having a long, happy, and accomplished career in statistics. However, continued learning is challenging given the day-to-day demands of most statistical jobs. Therefore, unless we plan efficient learning opportunities, little learning will take place. Although this book focused on the components essential to success in statistics other than technical acumen, improving technical acumen, in addition to the components covered in this book, through on-the-job learning is critical.

Yearly performance plans should include learning objectives along with a plan for how these objectives will be met. Although some learning opportunities may exist through attending conferences, seminars, lunch and learns, etc., it is also important to seek opportunities to learn as part of regularly assigned duties. These on-the-job learning opportunities are beneficial to both individual workers and the organization. The extra diligence in delivering assigned work helps ensure that work is of high quality, which benefits the organization and looks good on the individual's performance record, while also advancing the technical skills of the individual worker. Sharing key learnings with the organization also advances the individual's communication skills and increases institutional knowledge, which is good for the organization.

Although it may be challenging to find or make time for continued learning, utilizing the techniques to enhance productivity outlined in Chapter 2 make continued learning and maintaining work–life balance realistic.

When considering career development, longer-term planning, such as a three- to five-year plan is useful so that we can direct learning toward future goals and roles. Factors to consider in longer-term planning include breadth versus depth of knowledge, the benefits from coaches and mentors, and the benefits from coaching and mentoring, which may to some degree be influenced by the size of the organization.

Reading can be a time-efficient way to learn, especially about diverse topics, including some books for recreation to recharge. Read some books for career development including books in areas of interest to you and books in areas where you want to strengthen or broaden your skills.

IV

Other Perspectives

Preface to Part IV

PART IV INCLUDES CHAPTERS contributed by Christy Chuang-Stein, Marc Buyse, and Geert Molenberghs. Their perspectives on careers and career development help to broaden and reinforce the ideas from the earlier parts of the book. As in earlier chapters, you will not find many specific answers to specific questions in their chapters because individual situations can be idiosyncratic. However, the diversity in perspectives brought by these experienced and respected statisticians is important because it allows us to see how the basic principles covered in earlier chapters were applied by statisticians who built long, happy, and successful careers.

In Chapter 12, Christy Chuang-Stein shares her career journey in statistics and what she learned during that journey. Because many adult characteristics can be traced back to our upbringing, her story begins with her childhood. Regarding learnings, Christy focuses on what she experienced as a first-generation Chinese American, as a female statistician, and as someone whose English-speaking ability was limited when first arriving in the United States. Most of the chapter reflects her self-examination as a statistician in the pharmaceutical industry, but also reflects observations of her fellow statisticians.

In Chapter 13 Marc Buyse shares what he describes as his pseudo-random walk through life in statistics. Marc's path, like Geert's, Christy's, and mine, makes clear the unpredictability in careers, the unexpected twists and turns. Marc discusses key points raised in earlier chapters through the events that shaped

DOI: 10.1201/9781003334286-16

his career and life, including the chaotic COVID pandemic. Marc focuses on the idiosyncratic and personal nature of careers, including his own entertaining and inspirational anecdotes. Through Marc's experiences and reflections, we gain insight that can help shape our own careers.

Starting from his early education, in Chapter 14 Geert Molenberghs sketches the beginnings of his academic statistical career. Geert then discusses research and creativity, and how collaboration in its many forms can benefit that work. Geert's career journey illustrates how interactions with colleagues in settings somewhat different from our day-to-day work can benefit creativity. He shows us the crucial role mentoring and the learned societies play in academic life and in advancing science, along with providing an opportunity for an individual's growth. Geert's journey illustrates the diversity in skills required, many of which need to be learned on the job, over a career in statistics. The research, teaching, and consultancy required of many academic positions are already a diverse set of responsibilities, but many academic statisticians also take up roles in leadership and administration, which require still different skills. Geert's unique role in the COVID pandemic illustrates the challenges in communication with non-scientific audiences and how we can meet those challenges. Most of all Geert's career is a good illustration of work–life balance as he shows us the role that hobbies and sports activities have played in his life.

As you read these career journeys, look for the key points made in earlier chapters. For example, look for ways like those described in Chapter 2 for how these statisticians were so productive over prolonged periods; look for the strong working collaborations and friendships that fueled their careers; note how these relationships often began in informal settings and led to networks not unlike the 'coffee house networks' described in Chapter 3; and look for how they honed communication skills and how they developed critical thinking and leadership skills. Most of all, look for the ways in which they continued to learn on the job.

CHAPTER 12

Perspectives on Careers

Christy Chuang-Stein

ABSTRACT

In this chapter, Christy Chuang-Stein shares her career journey in statistics and what she learned during that journey. Because many adult characteristics can be traced back to our upbringing, her story begins with her childhood. Regarding learnings, Christy focuses on what she experienced as a first-generation Chinese American, as a female statistician, and as someone whose English-speaking ability was limited when first arriving in the United States. Most of the chapter reflects her self-examination as a statistician in the pharmaceutical industry, but also reflects observations of her fellow statisticians.

12.1 INTRODUCTION

A few months ago, I asked my son whether he remembered the good-better-best poem I shared with him when he was growing

DOI: 10.1201/9781003334286-17

up. He gave me his you-should-know-the-answer look and said, 'How could I not remember? You said it so many times!' The poem goes like this:

> Good, better, best;
> Never let it rest;
> Till good is better,
> And better best.

The poem sums up my drive for excellence in a foreign land, as a first-generation Chinese American statistician and as a female professional. Growing up with supportive patients (and yes, a demanding mother), the drive for excellence was ingrained in me from an early age. I was told that excellence needed to be a habit, not just a one-time act. It was only many years later, well into adulthood, that I became fully aware of how the pursuit of excellence shaped me and my relationships with those close to me, both professionally and personally.

As background to my story, the following are key positions in my professional career, listed chronologically.

- Assistant Professor of Oncology in Biostatistics, and of Statistics, University of Rochester

- Director, Clinical Biostatistics I, The Upjohn Company

- Senior Director, Head of Kalamazoo Biostatistics and Programming, Pharmacia Corporation

- Senior Director, Statistical Research and Consulting Center (SRCC), Pfizer

- Executive Director, Midwest Site Head of Statistics, Pfizer

- Executive Director, Midwest Site Head of Statistics, Head of SRCC, Pfizer

- Vice President, Head of SRCC, Pfizer

Some of these roles were a natural career progression while others were changes in response to an evolving environment. To understand my journey and what I learned along the way, it is best to start at the very beginning.

12.2 MY STORY

I grew up in Hualien, which is an eastern seaport in Taiwan. Despite its isolation, I loved growing up there because the sky was usually blue and the air was always fresh. In the early 1970s, Hualien was quite isolated, reachable from the capital city Taipei in the north via an eight-plus hour bus ride through a treacherous mountain road. The road was one-way traffic only, with a half-way transfer area where the traffic switched directions at noon.

My parents came to Hualien in the late 1940s. They came from different parts of China, drawn by teaching positions. They had their college degrees, but not much else. They met in Hualien, got married, and raised a family. My parents had four children, of which I am the third. We lived close to the Pacific Ocean. We could watch the sunrise over the ocean on most mornings. There was no TV in our house until the year I left for college. On summer nights, the sounds of waves crashing onto the rugged shore together with songs of frogs and crickets were our constant companions (Figure 12.1).

With no TV, every school night was homework night. My parents made it clear that the best way to secure our future was through education and hard work since they had no assets to pass onto us. Securing a future through education was the prevailing wisdom in China in my parents' generation. Since both of my parents had a college degree, they expected their kids to go beyond the four-year college education. The next generation needed to be better than the previous one, they often told us.

My mother taught math and history. My father taught biology and food processing. I remember doing math lessons with my mother in the summer while other kids were playing outside.

FIGURE 12.1   Picture of Christy with her parents and three siblings.

Yes, I resented the extra lessons at the time, but those lessons served me well. I became very good at math and ended up majoring in math in college.

During my junior year in college, math became too abstract for me. I wanted to learn how math could be applied to solve real-world problems. I took a year-long course on statistics during my senior year and I loved it! This led me to the School of Statistics at the University of Minnesota. In my generation, the United States was the destination of choice for most students who wanted to go to graduate schools. There were very few PhD programs offered by universities in Taiwan in the 1970s. My graduate work was mostly theoretical, supplemented by a few courses in biometry and applied statistics. I completed my PhD in 1980.

My first job after graduate school was a joint appointment at the University of Rochester in Rochester, New York. Half of my time was spent at the University's Cancer Center and the other half was devoted to teaching. My work at the Cancer Center made me realize that I wanted to spend all my time on biomedical research. In 1985 I joined The Upjohn Company in Kalamazoo Michigan and began my 30-year full-time employment in the pharmaceutical industry, a career where every day was an opportunity to work on a product that could save someone's life.

The pharmaceutical industry experienced huge growth during the last two decades of the 20th century. The growth kicked off numerous mergers and acquisitions. The Upjohn Company became Pharmacia and Upjohn, and then Pharmacia through two major mergers. I became the Head of Biostatistics and Programming at the Kalamazoo site. In October 2002, Pfizer announced its plan to acquire Pharmacia. The acquisition was finalized in early 2003, when it was also announced that all clinical research and development activities in Kalamazoo would be closed. Suddenly, I was facing a major decision about my job. I could find a new job in another company or accept a technical role located at Pfizer's Ann Arbor site. Either choice would mean starting new work relationships in a new organization.

In the end, the decision was based on family consideration. I joined Pfizer as a member of its Statistical Research and Consulting Center (SRCC) in July of 2003 and began my weekly commute between Kalamazoo and Ann Arbor (about 100 miles one way). A month later, my son started his freshman year at a public high school, by taking his science courses at the famed Math and Science Center in Kalamazoo with students from other high schools in the area. While the SRCC offered excellent opportunities to work on interesting projects and I enjoyed making new friends within Pfizer, the change was difficult emotionally because of the weekdays spent in Ann Arbor away from my family (Figure 12.2).

FIGURE 12.2 Christy with her colleagues in the Statistical Research and Consulting Center at Pfizer in 2004.

For two and half years, I was a technical contributor with no administrative responsibilities within the SRCC. I began to expand my external networks during that time. I enjoyed engaging in multiple professional societies, including associations focusing primarily on product development. I also started numerous research collaborations with colleagues across the pharmaceutical industry (Figure 12.3).

In February of 2006, I became the Midwest Site Head of Statistics. Nine months later, I took on the dual role as the Head of the Statistical Research and Consulting Center. My calendar suddenly became very crowded and time management became a top priority.

During a trip to Connecticut for a Pfizer statistical leadership meeting in early 2007, I attended an urgent group teleconference called by the Head of the Ann Arbor site. I learned that the company had decided to shut down the entire Ann Arbor site. I was stunned. The closure was going to affect thousands of company employees in Ann Arbor, many of whom had relocated to Ann Arbor from Kalamazoo just four years earlier.

FIGURE 12.3 Christy attending the first pharmaceutical industry Biostatistics and Data Management Leaders' summit in 2003.

I faced the same question as I had four years prior – do I stay or do I leave? My son would be heading off to college in the fall. My husband and I would be empty nesters. Even though my husband had retired, I could not imagine asking him to leave Kalamazoo because he loved the quality of life there. One possibility was to seek a position in the Biostatistics Department at the University of Michigan in Ann Arbor. Another possibility was to become an independent consultant, working from our Kalamazoo home.

In the end, my supervisor made the decision easy for me by offering to let me work from home as the Head of the Statistical Research and Consulting Center. So, in June 2007, I began working from home, which I did until I retired from Pfizer in July 2015. During the eight years of working from home, I traveled extensively to company sites to connect and to consult. I attended many conferences to meet with other statisticians. Working from home also meant summer days in Kalamazoo and winter days in

Florida. I was going to take full advantage of the working-from-home arrangement.

I had opportunities to take on more managerial responsibilities during the last eight years of my career at Pfizer. The opportunities would require me to station at a Pfizer site. I decided not to pursue those opportunities. I was content to spend my remaining time at Pfizer as an influential consultant and a supportive manager to the small group of senior consultants in the SRCC. I continued my extensive outreach activities with several professional societies, especially the American Statistical Association (Figure 12.4).

As of this writing, I am working as an independent statistical consultant. I am also mentoring several statisticians. I listen to their aspirations, take pride in their successes, and help them with their problems. I am finally able to achieve the work–life balance that had eluded me during most of my professional life.

An overarching lesson I have learned in my career is that our careers, like our lives, consist of a series of choices. We may not

FIGURE 12.4 Christy receiving the Founders Award from the President of the American Statistical Association in 2012.

always be able to control the outcomes of those choices, but we can control how we respond to the outcomes. Many of us have heard the advice to make lemonade when life gives us lemons. A positive attitude, coupled with an ability to get going when the going gets rough, has helped me build a long career with a strong sense of purpose and a deep feeling of fulfillment.

With the above as background, I will reflect on some of the points made in earlier chapters.

12.3 REFLECTIONS

12.3.1 Reflections on Chapter 1: Introduction

Your relationship with yourself sets the tone for every other relationship you have.

– ROBERT HOLDEN

Chapter 1 used a car as a model to describe the many parts of a statistical career, including statistical acumen, the ability to work productively, prioritization, communication skills, critical thinking, relationship skills, influence, leadership, career development, and continued learning. These skills are indeed critical to a successful and happy career. Most of these skills are what I call intellectual skills. In this section, I will focus on one aspect of relationship skills – the need to form a good emotional relationship with ourselves.

Leo Buscaglia once said, 'To love others you must first love yourself.' Self-love here does not mean narcissistic, excess self-absorption, but rather a healthy awareness of one's worth. While some scholars debate the necessity of self-love before the ability to love others, the statement nevertheless speaks to the importance of cultivating a positive attitude toward oneself.

Chapter 1 discusses the need for a defined purpose in life. What if the defined purpose is to get more of everything? I realized, late in my life, how the good-better-best culture under which I grew up has shaped me. My mother had high expectations for all her kids. She rarely praised us in our presence. She reminded us on many occasions that we needed to do more. As a result, I had a hard time figuring out what 'best' looked like for me. I struggled with questions like – was I good enough? Would others be happy with what I had to offer and consider my contributions valuable?

Interestingly, I am not alone in my uncertainty about self-worth. I have spoken with several first-generation statisticians who grew up with mothers like mine. Despite being professionally successful, they also have an inner voice questioning their own worth.

What can we do to overcome this self-doubt? Clearly, keeping our heads down and doing excellent work alone is not enough. I have learned that self-love is a good starting point. We need to make a covenant with ourselves to celebrate our achievements and stand up for what we rightfully deserve. The latter helps build self-esteem and self-confidence, which are essential to self-respect and happiness. But, how do we do this? I had asked friends whom I trusted to provide me with timely and objective feedback at different stages of my career. Over time, I learned not to feel defensive about suggestions and to accept compliments graciously. Through practice, I realized that I was worthy of that beautiful bright red windbreaker jacket in a fancy Nike store that was not on sale!

I can easily understand why parents love the good-better-best mantra. Unfortunately, following this motto mindlessly can lead to an endless pursuit of achievements, never being satisfied, and always wanting more. This overzealous approach will not result in happiness. As noted in Chapter 1, we should not focus so much on the end results that we forget to enjoy the trip to get to the end.

My son is a very good tennis player. He played varsity tennis in high school and in college. He told me that he enjoyed the training and enjoyed playing in tournaments. He gave his best on the court and made friends with his opponents off the court regardless of the outcome of a match. He knew that in big tournaments he would get beat as he moved up the brackets. But he loved the act of competition and the motion of going through one match after another. He has moved on from tennis to other life pursuits. His attitude about enjoying the journey has made him a balanced and generally happy person. In this regard, he is more his father's son than his mother's son.

The best gift we can give to ourselves is to take care of ourselves and to keep the only body and mind we have in good health. We should get into the habit of doing something for ourselves every day, even if just to take a short break for our own enjoyment, as Chapter 1 suggests. This helps build our sense of self-wellness, which ultimately contributes to the building of healthy relationships with others.

Before you can win, you have to believe you are worthy.

– MIKE DITKA

12.3.2 Reflections on Productivity, Prioritization, and Work–Life Balance

The main thing is to keep the main thing the main thing.

– STEPHEN COVEY

The early part of Chapter 2 discusses strategies to get the right (useful) work done efficiently so that we have time for the more important things in life. This reminds me of a common tendency to take care of the easier tasks first and leave time for the harder and perhaps more important tasks later. Have you noticed that many meetings are run this way? What happened at these meetings? The meetings often ran out of time to adequately tackle the more important problems at the end. At the personal level, the approach of doing the easy things first frequently leaves the more important tasks to a time when we have less energy and are less clear-headed. So, the effort to clear easier items on our plate first may not have worked out the way we intended.

I fell into this trap regularly in my life. While I have arranged to protect times when I could think clearly about the most important issues, I have not always been successful in resisting the temptation to clear the easier tasks first during the protected times. For one thing, clearing the easier tasks first made me feel good because I had managed to reduce my to-do lists substantially. So, knowing what is truly important and sticking to it is the first step to prioritizing. This includes not letting the noise of urgency create the illusion of importance.

President Lincoln once said, 'Give me six hours to chop down a tree and I will spend the first four sharpening the ax.' While a sharpened ax is necessary for a lumberjack to chop down a tree efficiently, it alone is not enough. The lumberjack must be physically and mentally fit to perform the task. Similarly, we can nurture our mental and physical health while we are developing other skills for success.

Chapter 2 talks about family, hobbies, and friendships that could help us lead a long, happy, and accomplished career. My husband has several hobbies. I, on the other hand, had none when I was working full-time. I used to brag about the fact that I did not need a hobby to offset job-related stress because I loved my job very much. Besides, who had time for hobbies? Certainly, not me. After I retired, I picked up gardening and biking. I have come to realize that hobbies are extremely effective in giving our mind a break and refreshing our thinking. Time spent on hobbies makes us more focused and productive when we return to work. I am not surprised to learn that many of Craig's great ideas first surfaced while playing fetch with Maggie.

Strategically planned time-off does not reduce our work output. On the contrary, it can make us happier and more eager to engage when we return (Figure 12.5).

Work–life balance is hard for all professionals with a family, especially so for professional women. I am no exception. Women tend to want to take care of everything and everyone, and we feel guilty if we can't do it all. I had a female colleague who felt compelled to change bathroom towels every day and bedsheets every couple of days. Imagine the amount of laundry she had to do? Why, I asked her? Part of this may have originated from

FIGURE 12.5   Christy enjoying riding her bike in Florida.

the time when women were homemakers. There are still many people who consider a woman's place to be at home, raising kids and supporting her husband.

Women should be the first ones to reject the notion that if we want to have a career outside the house, we need to be able to handle all the work inside the house ourselves first. Equally, we should reject the notion that we need to be a superwoman to win the respect of others. Nevertheless, we need to work with our partners to prioritize household chores and child-raising responsibilities. And we need to arrange for outside help to fill in the gaps. For example, we can hire helpers to clean our homes; we can do take-out from restaurants; and we can shop online. The help provided by others may not meet our high standards, but is it really that important that the floors are spotless in the first place? If we are willing to compromise on the less important things, we will have more time for the activities that should not be compromised.

I remember a time when our son was young and both my husband and I were working full-time. Before going on a business trip, I would make sure that the refrigerator was stocked and the laundry baskets were empty. When I returned home, I found that my husband and our son barely touched the food I prepared. My husband took our son to play golf. They tried out new restaurants. They even made a pizza from scratch. They did perfectly fine on their own. So why did I feel compelled to do all the prep work before my trips? Was it because I did not trust my husband, or because I wanted to feel needed? My husband didn't do similar prep work before he went on business trips. He trusted that I could take care of whatever needed to be done when he was away. I knew this because he told me so. In retrospect, I brought a lot of extra work – and stress – on myself unnecessarily.

There is no glory in being a superwoman. We only run ourselves ragged. Life and careers are marathons. We need to pace ourselves for the long game. There is no reward for excessive self-sacrifice either. A woman who knows her worth will ask for help because asking for help is not a sign of weakness, but a sign of strength to leverage the power of others. The asking is not limited to domestic tasks. It applies to our professional lives also.

> We need to do a better job of putting ourselves higher on our own 'To Do List.'
>
> – MICHELLE OBAMA

12.3.3 Reflections on Creativity and Innovation

> If you always do what you always did, you will always get what you always got.
>
> – ALBERT EINSTEIN

Organizations benefit from innovations that are well-integrated into the existing systems and embraced by their members. While this seems obvious, launching an innovation is often not straightforward. For one thing, innovation means change and many people don't like change. We may fear changes because changes disrupt the status quo. We tend to be comfortable with the status quo, even if that status is far from optimal. The following example illustrates these points.

Between 2008 and the time I retired from Pfizer in 2015, I chaired a multi-disciplinary team called Safety First. The team consisted of clinicians, safety risk leads, data managers, programmers, information technology specialists, and statisticians. The team helped set Pfizer policies on safety data review. Another critical action for the team was to develop a more efficient review tool.

At the time of the tool overhaul, reviews were done using line listings of adverse events produced in the portable document format from the company's clinical trial database and its safety database. A few basic tables summarizing event counts went with the listings. The reports did not allow reviewers to select subgroups for additional examination. After numerous focus group discussions, Safety First decided to develop a tool that offered graphical data presentations and allowed interactive data queries. The platform chosen was Spotfire`. The decision required approval from the leadership in the clinical and safety organizations because licenses for Spotfire` needed to be purchased and development resources needed to be allocated.

Safety First identified two senior leaders, one in the clinical organization and one in the safety organization, as sponsors of the tool. Once approval was received, development began. The input was sought from many future users during a year-long development effort. Once a prototype was developed, weekly training was offered by the developers as lunch-and-learn events. Figures like that in Figure 12.6 were pre-populated within the system. Training to a study team or to a department could be

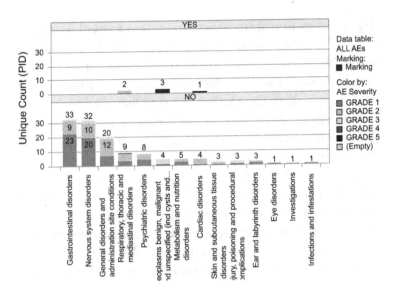

FIGURE 12.6   A standard graph produced by a new safety data review tool.

arranged. Safety First set a timetable for retiring the old reports of listings.

We were confident that the users would welcome the new tool with open arms and therefore expected rapid adoption. We were surprised to find that this was not the case. Clearly, our assumption that an excellent tool could speak for itself was wrong. We failed to consider the emotional implications of adopting new technology. We needed to do more to convince those who were wary of the new tool.

We took several steps to promote the tool. First, we enlisted early adopters to be the primary trainers. The early adopters showed their colleagues how, through a series of simple clicks, they could perform their reviews more efficiently. The early adopters showed how saved commands could produce a set of output on demand. Next, we asked our senior sponsors to share their expectations concerning the use of the new tool throughout their organizations. Third, we invited our user champions to present at the Global Education Forum to showcase the tool, as often as the forum organizers allowed us. Fourth, we asked members of Safety First to engage in personal outreach endeavors within their spheres of influence. The concerted efforts by many, coupled with top-down leadership and bottom-up influence, were ultimately successful in moving the organization to the new tool by the target timeline.

The experience taught me the importance of building strong connections with the community impacted by innovation. The connections need to be based on compassion and understanding. After all, it is people who make implementation of a new technology a success or a failure. One can't socialize a new process if there is no society standing ready to accept it. Coming up with a good idea is only the first step. Turning a good idea into a reality – that is what it really counts in the real world.

> For good ideas and true innovation, you need human interaction, conflict, argument, debate.
>
> – MARGARET HEFFERNAN

12.3.4 Reflections on Communication and Presentation Skills

> Communication works for those who work at it.
>
> – JOHN POWELL

I once wrote about a YouTube video (Biopharm Report, 2017). In the video, a blind man sat on a mat in front of a building. Next to him was a cardboard sign saying 'I'M BLIND. PLEASE HELP ME.' Occasionally a passerby dropped a coin on his mat. A woman walked by and saw the sign. She flipped the cardboard around and wrote a message on the other side. The blind man felt her shoes to get to know her. After the sign change, passersby started dropping handfuls of coins on his mat. At the end of the day, the woman returned and stood in front of the man. By the feel of the shoes, the man knew it was the same woman from the morning. He asked what she had done to his sign. She said,

'I wrote the same, but used different words.' The words she used were 'IT IS A BEAUTIFUL DAY and I CAN'T SEE IT.' The video clip ended with the message 'Words Matter!'

For most of us who came to attend graduate schools in the United States from East Asia, English is neither our first nor our second language. I remember the first time I held a recitation class as a graduate teaching assistant at the University of Minnesota. I was very nervous. I was worried that I would not be able to understand my students' questions. I was worried that I would not be able to formulate coherent answers in English fast enough. I was worried that my students would not be able to understand me because of my accent. I knew I needed help with my English, and I needed it fast. From that point on, I spent as much time as possible with my American officemates to improve my English. Few people knew this – I often read English newspapers out loud to practice my pronunciation during my first few years in the United States.

Today, more than 45 years later, I no longer worry about the basics of the English language. I will always speak English with an accent, but I am not self-conscious about it. There has been a growing acceptance of foreign accents, if the accent is not so thick as to hinder comprehension. I learned to speak English clearly with a good pace and good eye contact. But as noted in Chapter 4, good communication takes hard work.

On the writing part, I will always have a blind spot for the articles ('the' and 'a') and singular versus plural forms. Fortunately, I have a long-time collaborator who is British. He is also a good friend. For important writings, I ask him to review my draft and make necessary adjustments. I know my limits and am happy to ask for help. I have been fortunate to have received much editorial help in my career.

Over time, the focus of my communication shifted from grammar and clarity of speech to the effectiveness of the communication. As Lee Iacocca once said, 'You can have brilliant ideas; but if you can't get them across, your ideas won't get you anywhere.' I learned from experts that I could improve my communication, both written and oral, by paying attention to tone. For example, when reviewing drafts prepared for a situation, I'd consider whether I was using the right tone for that situation. Was I showing due respect and caring with the words I chose?

I learned the lesson of thinking twice before hitting the send-key the hard way. In 1987, I published a paper on a titration study in the journal *Statistics in Medicine*. A few years later, a statistician colleague sent a memo to senior leaders on titration studies and copied me. The memo contained some questionable arguments. It was obvious that my colleague never read my paper. I sent the same set of senior leaders a follow-up memo pointing out a better way to analyze a titration study. You guessed it! My colleague did not take my memo well. Since then, I have learned that private persuasion is usually more effective than public corrections.

On those occasions when we challenge others in public settings, we need to do it in a respectful manner. We should give others the benefit of doubt, assume good intentions, and be mindful that we don't know everything. Questions like 'I could have got this wrong, but as I see it...' and 'What do you think would happen if we did X instead of Y?' are effective ways to get others to think about our ideas. And, in the end, they will be more persuasive than statements like 'That's not right' or 'That plan won't work.'

Giving presentations is a regular part of life for many professionals. A presentation is a special form of communication. Presentations are typically more structured than general conversations and are usually theme-based. Chapter 4 noted that good presentations require preparation. But a presentation is not necessarily good simply because we spent a lot of time preparing for it. Personally, I have found the following seven tips to be especially helpful. These points distill much of the advice in Chapter 4 into a concise list. I learned the sixth tip from President J. F. Kennedy and Martin Luther King, Jr, who were among the best orators in modern US history (Figure 12.7).

FIGURE 12.7 Christy speaking at a plenary session at the First International Symposium on Biopharmaceutical Statistics in 2008.

1. Write out what I plan to say during the first three minutes. An excellent opener not only catches the audience's attention early, it also builds confidence and steadies the nerves.

2. Verbally rehearse the opener until I get it right, no stuttering and no tripping over the words.

3. Avoid using words that are hard for me to pronounce.

4. Keep the presentation fun with the strategic use of visual aids. Make sure that humor is not done at anyone's expense – but mine.

5. Put fresh energy and enthusiasm into each delivery – even if I have given a similar presentation many times.

6. Use different tones to emphasize the message and repeat the main message throughout the presentation. Chapter 4 describes through lines in presentations. The through line is often a main message that gets repeated in the talk.

7. Always plan for the presentation to be a couple minutes shorter than the time limit. Never exceed the time limit.
All the great speakers were bad speakers at first.

– RALPH WALDO EMERSON

12.3.5 Reflections on Critical Thinking and Decision under Uncertainty – Individual Factors

Nothing would be done at all if we waited until we could do it so well that no one could find fault with it.

– JOHN NEWMAN

Chapter 5 illustrates System 1 and System 2 thinking in Figure 5.1. Statisticians, guided by our quantitative training, tend to gravitate towards System 2. The hypothesis–testing framework, which is the basis for statistical inference in the frequentist world, requires the assessment of the probability of making erroneous decisions. Because of our expertise in risk assessment, we are often called upon by leaders in our organizations to ensure that nothing important is amiss before making a data-based decision.

We pride ourselves as gatekeepers of truth. We consider our voices to be voices of reason. We provide the needed balance to overly optimistic projections in a high-risk-high-reward business environment. Using tools and metrics such as the probability of success or assurance, we can objectively demonstrate to our leaders the best estimates of success, and to uncover reasons why the success rate for phase 3 programs is often not as high as desired or envisioned.

A challenge for statisticians with deeply ingrained System 2 thinking is to effectively communicate the outcome of our deliberations. If we share all that transpired in the thinking process, we would likely confuse the listeners. Our tendency to be thorough (which is a virtue by itself) could result in our losing the attention of our audience and having our deliberations dismissed. If we are reluctant to commit to a definitive answer when a definitive answer is called for, we would look indecisive.

The desire to cover all bases and the fear of making a mistake can compromise our effectiveness, especially in team meetings. What can we do to overcome our cautious nature and not let our training become our crutches? In my opinion, we need to learn to arrive at a position after balancing all the pertinent facts, assumptions, and opinions. If a reasonable position is not attainable, we need to learn to relay this conclusion and explain it in simple language. If we want to be a leader, we need to have the courage to stake a position after careful balancing of all the pros and cons. If someone else offers the same position as we are about to offer, we need to openly voice our support.

Offering our opinions publicly requires courage. At times, we may be concerned that no one would be interested in our ideas. Worse yet, we may be concerned that someone would make fun of our ideas, especially when we are still recovering from a prior negative experience.

There were times in my career when I sat in conference rooms with good ideas but failed to speak up. I told myself that there simply was not enough time to get into the details of the ideas. Besides, I needed more time to better formulate my oral delivery. I gave myself one excuse after another for not speaking up. I remember berating myself afterwards for missing a perfect opportunity to introduce a great idea and build consensus around it. When I finally overcame my hesitancy and submitted the idea in a follow-up memo, I was told that, while the idea sounded promising, the time for discussion was closed and a decision had already been made. Imagine how I felt then! And imagine, how other team members might feel if they knew that an opportunity was missed because of my hesitancy!

If I could do it all over again, I would learn to preface my ignorance and ask questions nevertheless. I would strive to speak up more and volunteer my opinions even if I am not sure about the best answer to the question under discussion. I would work harder earlier in my career to develop self-confidence and not let self-doubt deprive me of the opportunity to actively participate and contribute. I would openly embrace the final decision even if it is not the decision I originally supported. In the latter case, I will take pride in the fact that my participation helped others refine their original thinking and together we have reached a decision that is better than anyone of us could make alone. Similarly, I would try to remember the same efforts by others when my idea was adopted, that is, that others contributed too, and that the participation of others helped to make my idea, our idea, better.

Action cures fear; inaction creates terror.

– DOUGLAS HORTON

12.3.6 Reflections on Influence and Leadership

If you want to go fast, go alone.

If you want to go far, go together.

– AFRICAN PROVERB

There are many definitions for, and many types of leadership, as described in Chapter 7. When I think of leadership, I often gravitate to the simple characterization by Warren Bennis that leadership is the capacity to translate a vision into reality. This is accomplished by winning people to our way of thinking and joining in a cause. Under this characterization, leadership is an action – not a position.

There are two components to this characterization of leadership: (1) a well-articulated vision and (2) the ability to influence. Why is vision so important? Because vision serves as the North Star, it does not move and it guides us along the way. We cannot be a leader if we don't know where we are going. Many organizations have worked hard to arrive at their vision statements.

The second component of leadership is influence, to bring out actions in others. How can we move people who may not want to be moved? Chapter 7 covers many tactics developed by researchers. Personally, I have followed the seven simple rules given below. I can't claim authorship to any of them, but they are easy to remember and resonate with me.

1. Give sincere appreciation and honest feedback. There is an old saying – If we want to gather honey, we should not kick over the beehive. The most effective way to bring out the best in others is through appreciation and encouragement.

2. Arouse a strong desire in others to dream with us. Make others feel that they are part of the journey by incorporating their views and ideas, if possible.

3. Cultivate a positive and optimistic atmosphere. A can-do attitude is contagious and a force multiplier. A leader's attitude often determines the attitudes of the followers.

4. Always display integrity to earn trust. Someone once compared trust to fine China – it is very expensive to acquire but it is easy to break. We build trust with others each time we place integrity over image and truth over convenience. In addition to showing that we are trustworthy, we need to show that we are willing to believe in others' integrity to do the right thing, at possibly considerable risk to ourselves.

5. Act with transparency and share credit with others. Acknowledge publicly the contributions from others, no matter how small the contributions may be. President Truman once said, 'It is amazing what you can accomplish if you do not care who gets the credit.'

6. Be humble. C. S. Lewis said that true humility is not thinking less of ourselves, but thinking of ourselves less. Admit our mistakes openly and quickly. We are more likely to win

others' trust if we have the courage to admit our own mistakes than try to cover them up.

7. Finally, we need to lead by example. Great generals join their men and women on the frontlines in the heat of a battle. The ability to stride down to the end of a dark tunnel and light up a candle contributes greatly to the making of a leader.

Is charisma necessary for leadership? In a *New York Times* article on August 15, 2019, Bryan Clark cited Olivia Fox Cabane (the author of the book *The Charismatic Myth*) that charismatic behaviors could be boiled down to three pillars. They are presence, power, and warmth. Cabane explained the core quality of presence to be the ability to focus totally on someone we are interacting with. Power is the ability to break down self-imposed barriers and replace them with self-confidence and a strong belief in oneself. Warmth pertains to the vibe that signals kindness and acceptance of others, no matter who they are and what they are.

The article went on to give examples of people considered to be great leaders. It argued that they tend to belong to one of two groups: those who project charm through warmth, generosity, and empathy, the Gandhi type; and, those who exhibit great self-confidence and success, the Steve Jobs type. The article concluded that it is hard to be an effective leader if one did not have any of the three charismatic qualities discussed in the article. I was relieved to find that, in the opinion of the columnist, good-looking and smooth-talking were neither a necessary nor a sufficient condition for being an effective leader.

The behaviors described above take time to develop, and that development is a continuous improvement process. When we first learned to walk, we fell often. The falls did not deter us from trying again. We should apply the same spirit to developing our leadership skills. It takes many decades for a sequoia tree to grow

FIGURE 12.8 Christy standing in front of two giant trees at the Mt. Rainier National Park in 2009.

to the giant size that can awe and inspire. We should not expect our efforts to develop leadership skills to bear fruit overnight (Figure 12.8).

Let me now turn to statistical leadership. Statisticians have made great inroads during the past 30 years. We have learned to be proactive and innovative in study design, study conduct, data analysis, data interpretation, and data presentation. We take pride in being the best storytellers of data with an ability to unearth hidden messages.

A shining example of statistical leadership occurred in the 1990s during the human immunodeficiency virus (HIV) crisis. The first couple approvals of drugs to treat the primary infection of HIV were based on a clinical endpoint (e.g., death or opportunistic infections). With HIV spreading rapidly, there was an urgent need to speed up the testing and approval of new drugs. A central question emerged: Could a surrogate marker be identified to replace the clinical endpoint? To address this question, a Surrogate Marker Collaborative Group (SMCG) was formed with members from academia, government, and industry. The main action was to interrogate accumulating data at each member institution to address a set of questions. Data, prepared in a common format, were submitted to an academic member of the SMCG for meta-analyses. Statisticians took leadership roles within their individual organizations. Over time, the search for markers narrowed down to the potential role of viral load suppression as a replacement for the clinical endpoint. SMCG members presented their data on viral load suppression and its relationship to clinical outcome at an FDA Antiviral Drugs Advisory Committee meeting on July 14–15, 1997.

A similar meeting took place with the Committee for Medicinal Products for Human Use (CHMP) of the European Medicine Agency on September 22–23 of the same year. I presented data from the Pharmacia and Upjohn Company at the FDA meeting and participated in the CHMP discussions. Shortly after the meetings, new regulatory guidance was published by the FDA and the CHMP. The new guidance accepted viral load suppression as a primary endpoint for full approvals of HIV drugs. The joint efforts by many, especially with statisticians in leadership roles, resulted in an important change, to no longer require clinical outcome trials for full approvals. This led to expedited development and approvals of many HIV drugs in the last two decades.

Unfortunately, the HIV success story isn't the way things always turn out. Why do we statisticians sometimes struggle to have our voices heard, to have our contributions recognized, and to have a seat at the decision-making table? How could this be when we have so much to contribute?

A critical trait that we need to cultivate is the projection of power in the context of the *New York Times* article. As I discussed in my reflections on Chapter 5, statisticians were trained not to judge (or to judge with a lot of qualifying statements) until we have overwhelming evidence. Many of us feel that our job is to present data and let others draw their own conclusions. Even when we have all the data we will ever get, we may still be reluctant to take a position when that is what is required of us. Without a position, we are left behind and skipped over when senior management looks for leadership.

Taking a position means taking a risk that we may be wrong. It requires self-confidence to take bold actions in the face of risks. In a presentation to the leaders of the European Federation of Statisticians in the Pharmaceutical Industry (EFSPI) in celebration of EFSPI's tenth anniversary, I encouraged attendees to cultivate bold and entrepreneurial behaviors among their statisticians. I believe that such behaviors can help moderate statisticians' risk-averse inclinations and propel us to new opportunities (Figure 12.9).

When we are willing to take risks, what we are really saying is that – we believe in tomorrow, and we want to be part of it. If you were a ship, would you prefer staying in a safe harbor or would you prefer venturing out into the open waters for exciting adventures?

> A great leader is not a searcher for consensus, but a molder of consensus.
>
> – MARTIN LUTHER KING JR

12.3.7 Reflection on Relationships at Work

> We can improve our relationships with others by leaps and bounds if we become encouragers instead of critics.
>
> – JOYCE MEYER

We need each other, plain and simple. Remember the movie *Cast Away*? The FedEx cargo plane to Malaysia that Chuck Noland

FIGURE 12.9 Christy with other leaders at a meeting celebrating the tenth anniversary of the European Federation of Statisticians in the Pharmaceutical Industry in 2019.

(played by Tom Hanks) flew on crashed into the Pacific Ocean during a storm. Washed up on an uninhabited island with an inflatable life raft, Chuck was the sole survivor. He coped with his forced isolation by inventing an imaginary friend in a volley-ball which he named Wilson. Chuck talked to Wilson regularly and grieved over the loss of Wilson when Wilson floated away atop a raft he constructed. He kept a FedEx package with a pair of painted golden angel wings throughout his ordeal, holding onto the hope that he would return the package to the sender one day. That hope helped sustain him and ultimately helped save his life.

Many of the adjectives describing our character are defined in terms of our relationships with others. For example, we cannot be kind or loving if there is no one for us to be kind or loving to. When we first learned to speak, we imitated those around us. As we grew older, we identified idols and emulated them. As professionals, we picked our role models for their inspiring qualities and modeled our behaviors after theirs.

Good relationships enrich life. This is also true for relationships at work since we spend so much time at work. Besides, work offers us opportunities for instant friendships by placing us on the same project. Work is much more fun when we look forward to seeing our coworkers and having lunch with them. Some of our coworkers became our confidants as the friendships deepened. Among the things that I missed the most upon retiring from Pfizer in 2015 was the friendship I had built over the years. Many of my coworkers became like family members to me (Figure 12.10).

Joyce Meyer said that we could greatly improve our relationships with others if we could be encouragers instead of critics. I have often wondered why it is hard for many statisticians to praise their colleagues openly. Is it because our training focused us on finding fault? Is it because we are afraid that praise would lead others to overlook potential pitfalls and go down a wrong

FIGURE 12.10 The well-wish cake at Christy's retirement party at Pfizer in 2015.

path? Does the desire to be right lead us to focus on the short-coming of a new proposal even though the current approach has more problems? In doing so, are we letting perfect become the enemy of the good?

It took me many years to realize the power of public praise. I did not appreciate how much public praise meant until I was praised publicly by Professor Gary Koch for my work on safety data analysis in the 1990s. The writer J. R. R. Tolkien once said, 'The praise of the praiseworthy is above all rewards.' I listed giving sincere appreciation and honest feedback as a top strategy for leadership and influencing others. There is a Yiddish phrase – praise the young and they will blossom. I wish that someone had introduced this phrase to my mother when I grew up. It was good that my mother had high expectations of her children. However, praise is the other side of the relationship equation, and things work best when the equation is balanced.

Occasionally, we are put in a position to compete with our colleagues. This could be a deliberate act by a competitive leader or the unfortunate consequence of a curve-based performance grading system. There may be less incentive to make our colleagues look good in such situations. Even so, we need to give our colleagues the compliments they deserve. Our show of respect to others and our gesture of fair play will not go unnoticed. Someone once said, 'Speak kind words and you will hear kind echoes.'

I want to make a special plea for female statisticians to support our female colleagues. The friendships built between female colleagues often extend to other areas in life because of the common struggles we face. We need to help lift each other up, and not push each other down due to some ill-conceived notion of competition.

What should we do when an environment became unhealthy or even toxic? Instead of becoming an unhappy contributor, it is usually best to look for new opportunities. The new opportunities do not necessarily have to be in the same segment of employment

or the same type of job. I have served on many award committees during my career. I have seen many outstanding statisticians who have chosen to leave the comfortable folds of statistics
and become prominent in another field. Some have moved into
administration and become the dean or provost of a college (yes,
even the president of a university). Some statisticians in the pharmaceutical industry have crossed over to other disciplines such
as regulatory affairs, safety, information technology, or project
management. I have seen them applying their well-honed logical thinking to their new positions and found great successes in
their new homes.

Coming together is a beginning. Keeping together is progress. Working together is success.

– AUTHOR UNKNOWN

12.3.8 Reflections on Career Planning and Continued Learning

Tomorrow belongs to people who prepare for it today.

– AFRICAN PROVERB

I have long believed that our career development is our responsibility, with hopefully the support from our managers and
loved ones. Our academic training gives us a foundation for
select technical skills. Once we start a job, on-the-job training
takes over. Chapter 10 shared the opinion of a hiring manager
that 90% of what we need to know would be learned on the
job. Chuang-Stein (1996) argued that unlike coursework in a
graduate program, on-the-job training does not have a clear
beginning nor a clear end. The journey takes us to a higher and
higher ground that allows us to see further and further into the
horizon.

My son, the tennis player, has settled down in Seattle with his
wife. They are both avid backcountry skiers. During a backcountry ski excursion, they climb up high with their skis and make

only one or two runs down the snowfield in a day. But the views they see, the views they share with us, are stunning. The climb was hard, they said. But the reward of great vistas has drawn them farther and farther into remote mountain passes where few have visited. For those of us who prefer summer hiking to winter skiing, we also enjoy the view as we climb high. On those occasions when we stop at a lookout spot, we are often amazed by how far we have traveled. Career development is like that long ascending trail (Figure 12.11).

Setbacks and disappointments are a natural part of our development. I received rejection letters for my papers throughout my career. Early in my career, I had to let the rejection letters sit for a few days before I could read the critiques. Even though I was advised to treat feedback as a gift, it was hard to read the critiques. Thanks to these critiques and accumulating experience, I learned to write better papers.

FIGURE 12.11   Christy's son going up towards the Glacier Peak in the state of Washington, May 2021.

I was nominated to run for the ASA president in 2013. Had I been elected I would have served as the association's president in 2015. I was disappointed that I did not win. But I directed my energy to the National Institute of Statistical Sciences (NISS) after the election and chaired its Affiliated Committee for four years. Together with other dedicated members, the committee kicked off several new programs for the institute and its affiliates. The work has earned me the Distinguished Service Award from NISS in 2020. It is often said that – when one door closes, another one usually opens. Helen Keller advised us not to look too long at the closed door because that would prevent us from seeing the door that had opened for us.

When we challenge ourselves with new responsibilities, we will often be out of our comfort zone. And we will make mistakes. How we handle mistakes and rebound from them will speak to our resilience. It is said that a wise person learned from his mistakes; a wiser one learned from others' mistakes; but the wisest person of all learned from others' successes. There are many learning opportunities in our continuous development journey as long as we are willing to keep our minds open.

Many years ago, I saw a picture which stayed in my mind to this day. That picture showed a cute little boy, perhaps eight years old. He was holding something and had a huge grin on his face as he was bursting with excitement. He couldn't wait to show others his prized possession. He had a live frog! To him, that frog was the most precious thing in his world at that moment. We need to continuously look for that frog in our career. We need to keep the fire of curiosity burning. The fire will help us learn and search for the next innovations. Like Steve Jobs – pursuing something that others have not even begun to imagine!

Excitement and stimulation may come from a career change. Chuang-Stein (2019) pointed out that statisticians working in the pharmaceutical industry are familiar with go/no-go milestone decisions. There are times when we face similar decisions in our career journeys. The timing of the decisions may vary among individuals. Yet, sooner or later, most of us will face questions like: Should I apply for a higher position within my organization or in a different organization? Will I be good at it? Will I enjoy the prospect of working in a new area? What if I fail in the new position? Will my reputation suffer irreparable damage if I fail in a new pursuit?

Interestingly, some statisticians find it difficult to make a career change. I wonder whether this has something to do with the statistical training I discussed earlier. Statisticians are good at calculating risks associated with wrong decisions, but we may not be as good at articulating gains from right decisions. We caution against making go-decisions when the evidence for moving forward is limited. This cautious tendency may make it hard to make a career change in the face of uncertainties.

I picked up the game of duplicate bridge after my retirement from Pfizer. My husband and I played regularly at our local bridge club before the COVID-19 pandemic. I often reviewed the hands played afterwards. I found that players who are willing to take a risk and bid for higher-level contracts tend to have higher scores in the long run. Could it be that life is like a series of bridge hands? One can play it safe and be good. But, could one become great if willing to take calculated risks? Data have suggested this to be the case.

So, next time, when you face a career decision, don't be afraid to take a risk. The decision to move forward may open doors to other opportunities that you have never imagined before. Be an entrepreneur and be bold in your decision. The change may seem daunting at the beginning, but it could also be energizing. Someone once told me, if you want to be a diamond, you need to be willing to be cut.

I do not believe things happen accidentally; I believe we earn them.

– MADELEINE ALBRIGHT

12.3.9 Reflections on Craig's List – Christy's Additions

Many of my friends know that I like quotes because quotes remind me of many facts in life with just a few words. I have shared many quotes in this chapter. I am adding eight more here.

Smile when picking up the phone. The caller will hear it in your voice.

> – AUTHOR UNKNOWN

Work is a slice of your life. It is not the entire pizza.

> – JACQUELYN MITCHARD

If you are patient in one moment of anger, you will escape a hundred days of sorrow.

> – AUTHOR UNKNOWN

The only limit to our realization of tomorrow will be our doubts of today.

> – FRANKLIN D. ROOSEVELT

Life begets life. Energy creates energy. It is by spending ourselves that one becomes rich.

> – SARAH BERNHARDT

I cannot imagine a person becoming a success who doesn't give this game of life everything he's got.

> – WATER CRONKITE

Should you shield the canyons from the windstorms, you would never see the beauty of their carvings.

> – ELISABETH KUBLER-ROSS

The man who goes farthest is generally the one who is willing to do and dare. The sure-thing boat never gets far from the shore.

> – DALE CARNEGIE

CHAPTER 13

Perspectives on Careers

Marc Buyse

ABSTRACT

In this chapter, Marc Buyse shares what he describes as his pseudo-random walk through life in statistics. Marc discusses key points raised in earlier chapters through the events that shaped his career and life, including the chaotic COVID pandemic. Having owned his own company, Marc brings an entrepreneur's perspective to careers. He discusses risk-taking and sensitive topics in statistics, such as data manipulation and fraud. Much of his focus is on the idiosyncratic and personal nature of careers, including his own entertaining and inspirational anecdotes. Through Marc's experiences and reflections, we gain insight that can help shape our own careers.

DOI: 10.1201/9781003334286-18

13.1 INTRODUCTION

This chapter is a pseudo-random walk through my life in statistics. I reflected on events that shaped my career and my life. This reflection is influenced by the chaotic period we are going through as I draft this chapter (the unprecedented COVID-19 pandemic, a war in Ukraine, and political unrest worldwide), which has led me to discuss the impact of sensitive topics such as lying or cheating on the statistical profession.

I wish I could think of specific recommendations for future generations, but it seems to me that the path of individuals through the circumstances of their life is highly personal and can hardly be guided by canned recommendations. I do believe, however, that having knowledge of general principles, such as those covered in earlier chapters of this book, provides the raw material from which good decisions can be made. And, to the point of this chapter, reading someone else's life story can help us in applying the principles and in shaping our habits, beliefs, and behaviors. Much of this chapter is anecdotal, but perhaps a few messages will emerge that some readers will find entertaining, if not inspirational.

13.1.1 First Course in Statistics – What a Bore!

Statisticians: twice as boring as accountants and half as rich.

STEPHEN SENN

Ask anyone how they remember their first course in statistics, the answer is almost guaranteed to be: '*I hated it!*' I am no exception. However, for most people, this first course in statistics is also their last, to their great relief. In this, I differ from the majority, fortunately. My first graduate studies were in engineering, and I was utterly bored by the only statistical course we had in our five-year curriculum. I did well in this course, but only because the exam was just as useless as the course itself: A formality as trivial and tedious as filling an administrative document required to get one's degree. The course was entitled 'Probability theory and statistical inference.' Looking back at the title of the course (which is the only thing I remember from it), there was no chance it could have interested me, let alone enthused me, at the time I took it. And today, with some in-depth knowledge of these topics, I wonder how anyone could be expected to cover them in 30 hours of undergraduate classes... That says a lot about how statistics is taught, or for that matter, how most other subjects are taught. But I'll get back to education later.

13.1.2 Second Course in Statistics – Love at Second Sight

The quiet statisticians have changed our world (…)

by changing the ways that we reason, experiment and form our opinions.

                              _   *IAN HACKING*

I completed my engineering degree in computer science and was offered a scholarship to complete my curriculum with an MBA at Cranfield University (United Kingdom). That degree too included a course in statistics called 'Business Statistics.' At least this title was comprehensible, but I was not thrilled to dive back into a topic that had left no pleasant or lasting trace in my memory. Yet what I least expected happened: I absolutely loved this course, perhaps in part because I did not sweat to understand it, but much more likely because it was taught by Tom Cass, an amazingly clear, entertaining, and charismatic professor.

The concepts that I had previously failed to grasp became obvious and, more importantly, fascinating. Tom Cass demonstrated, through concrete case studies, how statistics could be used to address complex problems, inform decision-making, understand relationships between observations, and so on and so forth. It opened a world of possibilities. To me, Tom's course was the pinnacle of the MBA program, the only one that touched on challenging concepts. It intrigued me and kept me thinking about why this odd branch of mathematics was so apt for addressing real-world problems. The impression stuck, and to this day I am awed by the fact that statistics goes far beyond describing the world around us: It helps to understand the world and, whenever possible, to control it.

13.1.3 Biostatistics – It's All in the 'Bio'

Let the dataset change your mindset.

HANS ROSLING (1948–2017)

A few months later, with two degrees in my pocket, I was ready to conquer the world. I had good degrees, but poor understanding of what I wanted to spend my life on. My father arranged a few interviews with select friends of his, so that I could see for myself what they were doing on a daily basis. For young graduates who are seeking a path forward, such interviews may not be useful to find out what they want to do; however, such interviews are incredibly useful to find out what they do *not* want to do.

I first met a friend of my father who had had a stellar career in a large multinational corporation. In my interview with him, on the top floor of a building whose layout emphasized hierarchy rather than efficiency, I expressed my overarching desire to work on interesting problems. He said that sooner or later I would have to make a choice between working on interesting problems or making money. On leaving his office, I knew I would spend my life proving him wrong.

My father also had friends in academia. He arranged for me to meet Professor Henri Tagnon (1911–2000), who had founded the European Organisation for Research and Treatment of Cancer (EORTC) (Meunier & van Oosterom, 2001). This interview was quite different: I met Professor Tagnon in a small, congested office at Institut Jules Bordet, the Belgian Cancer Center in Brussels. He did not bother to ask me what I wanted to do, but he told me about his own career, why he had left Belgium to work at the Harvard Medical School and Memorial Sloan-Kettering Cancer Center, and why he thought cancer research had to be funded and coordinated at the European level. Above all, he made a passionate plea for basing treatment decisions in cancer on data rather than dogma. On leaving his office, I knew I had found my path: I would join the EORTC. I would work on interesting problems and help find better treatments for patients. Making money had become a secondary concern. A few days later, I had traded the comforts of a busy office building for a deserted radiography room in the basement of Institut Jules Bordet. I embarked on an MSc degree in statistics and later a PhD in biostatistics. Statistics was a fascinating set of tools, but the 'bio' part of it had become my true passion.

13.2 REFLECTIONS

13.2.1 On Events That Matter

The beauty of life is in small details, not in big events.

JIM JARMUSCH

Statisticians will gladly accept that their life is largely shaped by random events. Of these, the most significant are, without a doubt, personal encounters. Which explains why the lockdowns during the COVID-19 pandemic were so painful for young people whose life still very much depends on such encounters. The two aspects of the biostatistical profession that I have thoroughly enjoyed are the conferences I could attend and the people I would meet at these conferences. Conferences are an unmatchable way to travel with a purpose. I don't mind sitting on a beach instead of attending all the conference sessions, but I do mind having to sit on a beach without any conference sessions to attend.

The first conference I attended was the inaugural conference of the International Society of Clinical Biostatistics (ISCB), an organization that had been single-handedly founded by Professor Maurice Staquet, my boss at the time (Duez et al., 2013). He had attracted the *fine fleur* of biostatistics at that time: David Cox, Norman Breslow, Edmund Gehan, Nathan Mantel, and other giants admired by my colleagues, but whose reputations I did not yet fully appreciate. Figures 13.1 and 13.2 provide two views of the inaugural session of this conference.

In Figure 13.1, Prof Maurice Staquet, ISCB Founder, sits in the first row in a white coat, with Richard Sylvester, then Director of Biostatistics at the EORTC Data Center, to his left. Readers may have fun spotting several illustrious biostatisticians of the time seated in this small auditorium.

This being the first conference of a nascent society, there were not many attendees. I was tasked to give a talk on behalf of the EORTC statistical department. I got much help from my colleagues (most already had a PhD, whereas I had barely completed my MSc), but even so, the contents of my talk were rather thin. I gave my presentation with more conviction than was called for, but at least I was not booed. As soon as I finished, Ed Gehan raised his hand seemingly to ask a question. I still remember the terror that suddenly overwhelmed me. I had not prepared for questions, and I was going to get grilled, I would have to leave the room in shame, and this presentation would mark the end of

FIGURE 13.1   The first meeting of the International Society for Clinical Biostatistics in Brussels, Belgium, 1979.

FIGURE 13.2   First meeting of ISCB (Brussels, 1979). Further illustrious biostatisticians of the time can be spotted in this picture. I sit in the third position from the left in the top row, seemingly wondering what I am doing in such an austere setting.

my career in biostatistics. To my surprise, Ed did not ask a question, he congratulated me for the interesting approach I had presented. I knew the compliment was not deserved, as I was acutely aware of the limitations of my talk. Regardless, Ed's gentle comment gave me the confidence I needed to go forward. Attending conferences and giving talks would become one of the great joys of my career.

13.2.2 On Kindness

The ideals which have lighted my way, and time after time, have given me new courage to face life cheerfully, have been kindness, beauty and truth.

ALBERT EINSTEIN (1879–1955)

I've often remembered the kindness of Ed Gehan's comment after my presentation. He could have been critical, and he would have gained nothing in the process, while I would have lost a lot – perhaps even the confidence I had in my ability to become a statistician. It is strange that society often equates kindness with weakness. Is our educational system at the root of this belief, with its emphasis on sanctions rather than on rewards?

It has taken me a long time to understand that being kind and rewarding to others is much more conducive to progress than being unsupportive and disparaging. Take, for instance, peer review. How many times have you had to review a paper of marginal interest – let alone one that is plain wrong? Yet the authors of these papers have spent a great deal of time on them, and as such, they deserve to be helped in improving their manuscript, regardless of how much improvement is called for (sometimes the task appears insurmountable, but no situation is as desperate as one thinks).

I have come to see peer reviews as acts of kindness and positive reinforcement, rather than sanctions against poor work and bad research. This is not to say I systematically recommend acceptance without revision; it merely means that I will carefully justify my opinion in as much detail as I can to help the authors improve their paper. And, even if the authors do not address my criticisms, I will politely insist on changes that I consider essential, but without making it a personal matter. I will always write my review as if my name was known to the authors (which is a sensible policy adopted by some journals), and as if their names were not known to me (again, a sensible policy adopted by some journals). It does improve the thoroughness with which I do my review – something authors expect, but do not always get. And I make sure that even if I suggest rejecting a paper, I do so with kindness and respect – something else authors expect, but do not always get! I will have more to say about peer review later.

I've had to work on being kind, but not very much. The main reason is probably that my father must have been the kindest man on the surface of the earth. At least I thought so – and I still do. Mind you, he could also lose his temper without warning. His bouts of rage were just scary enough to give me pause, but not violent enough to seriously harm anyone – like when he threw a slice of cream cake at my mother during a memorable Sunday brunch, making sure to miss her and to spoil the carpet instead. This formidable scene shaped my character better than any lengthy discourse: My father had given me proof that he was not perfect. I knew he had behaved badly, but I admired the fact that he had done so without being violent. He had made his point with force, if not with elegance.

That day, he taught me that in complicated situations, one can compromise on some principles, but not on others. Being kind and at the same time assertive has probably been the greatest asset I've been able to count on. I know I can get things done my way without unduly antagonizing others. I've often observed brilliant people lose an argument because they were unnecessarily aggressive about it. Brilliance does not excuse abrasiveness.

13.2.3 On Lying

Figures often beguile me, particularly when I have the arranging of them myself; in which case the remark attributed to Disraeli would often apply with justice and force:

'There are three kinds of lies: lies, damned lies, and statistics.'

MARK TWAIN (1835–1910)

We statisticians can only be antagonized by the infamous quote 'lies, damn lies, and statistics.' This quote looks silly and offensive. However, the short quote is taken out of its context, and it is well worth reading the full citation in Mark Twain's autobiography (Twain, 1906). In its totality, Twain's text is at once funny and to the point, which is not that statistics are worse than lies, but that statistics can easily be misused, either through incompetence (as in Mark Twain's case) or dishonesty (as in the cases implied by those who misquote him). The irony, of course, is that the very purpose of statistics, when well used, is to uncover the truth, given the available data, not to distort it – the exact opposite of Twain's quote. What we mean by 'the truth' must however be qualified to be comprehensible.

Let's first see what we mean by lies, a concept that is easier to grasp than 'the truth.' I learned early in my childhood that there are lies and lies. Like all kids, I knew there were lies *by omission*, and I reckoned it was my parents' privilege to decide whether to inform their children of something that might be important or relevant. In adult life, lies by omission are so pervasive that one tends to consider them the norm, and the only place where they play a prominent role is a court of justice. Life would likely not be manageable if we had to constantly make sure that we do not lie by omission.

The second type of lies I discovered was more of a shock: Santa Claus was not an old man with a large white beard who descended from heaven and went down our chimney to bring us presents. The huge disappointment of this discovery was followed by the uneasy feeling that parents could lie in other ways than by omission, albeit with the best of intentions (after all, this lie cost them dearly in monetary terms). At any rate, I realized that there exist *protective* lies.

In adult life, I encountered similar lies in my early career at Institut Jules Bordet: In those days (the late 1970s), patients were not always told about their diagnosis, and the word 'cancer' was carefully omitted from all communications. Institut Jules Bordet was called a 'Tumor Center,' not a 'Cancer Center' for the same reason. Not to share a diagnosis with a patient seems abhorrent today, but such was not the case in my country 40 years ago, and it is still common practice in many countries today, so it is fair to say that protective lies may be acceptable in some circumstances and social environments. I personally hope that the need for full transparency will not result in Santa Claus being demystified!

The last type of lies I discovered in my early childhood was far more interesting. My mother used them so abundantly that I could easily catch her in the positive act of lying to me. However, she went through great pains to explain that her lies were *useful* lies, and that it was perfectly fine to use them when needed. I soon realized she meant useful to her, for instance when she promised something and then forgot about it. If she could not remember a promise she had made, then she was no longer bound by it. Her logic seemed impeccable and left me speechless at the time. Thankfully, my father never used these so-called *useful* lies – though I am sure he must have lied a lot by omission, and he certainly never told us the truth about Santa Claus. My father always held the higher moral ground and that cast serious doubts on my mothers' theory of useful lies.

In fairness, it is now clear to me that my mother spent most of her time with her children, while my father spent very little time with us. As a result of this exposure bias, her need for useful lies may have been much more acute than his. We tend to reproduce our parents' behavior in adult life, and when it comes to lying, I must confess I've always been torn between the ideals set forth by my father and the pragmatism of my mother. However, when in doubt, my profession has helped me a great deal to make a clear-cut choice: Useful lies are to be banned without mercy. Statisticians are in the business of uncovering some hidden truth. It would be a professional misdemeanor to use statistics to lie.

13.2.4 On Statistics and Truth

My intention is not to prove that I was right but to find out whether I was right.

BERTOLT BRECHT (1898–1956), THE LIFE OF GALILEO

But what do we mean by 'some hidden truth'? The wonderful book *Statistics and Truth* (Rao, 1989) contains many insights into this matter. In statistics, we take a very narrow view of 'the truth' to make it a workable concept for our methods. In typical situations, we have interest in estimating an unknown parameter – say, the effect of a new treatment on a specific disease symptom or other patient-relevant outcome. The truth for us is the true value of this treatment effect. Unfortunately, we can never claim to know this treatment effect with certainty. Indeed, if we conduct an experiment – say, a randomized clinical trial – to estimate the treatment effect, the observed effect may differ from the true effect because of errors, which fall into two distinct categories: Systematic errors (bias) and random errors. This is stated in simple terms by the following equation, which was used by Sir Richard Peto of Oxford University to explain the essence of statistical estimation to collaborative groups of clinicians conducting trials in cardiovascular disease and cancer. I am pretty sure his equation illuminated their life (and mine) more than dozens of hours of tedious statistical classes.

observed treatment effect
= true treatment effect
+ systematic error (bias)
+ random error

Peto's equation was effective to convince clinicians of the need for randomized evidence (to eliminate systematic errors) and of the need for large numbers of observations (to reduce random errors) (Collins et al., 2020). The concepts of bias and randomness are so important and so obvious that they should be part of our basic education. Peto's equation could indeed be taught, with appropriate examples, to children of primary school age.

Estimation is alright but statisticians do not stop there: They also want to test hypotheses (except if they are Bayesian, which is fine by you and me, but not so straightforward for a new drug to gain regulatory approval). Hypothesis testing becomes second nature once you get used to it, but in fairness to clinicians, it is a rather contrived maneuver that starts by postulating that there is no treatment effect at all (the null hypothesis, which is the opposite of the situation of interest).

This starting point makes the statistical profession a rather unfriendly one in medical circles. Once data become available, statisticians analyze the data, and, more often than not, they conclude that the data are insufficient to claim that there is a treatment effect. This makes them doubly unfriendly. Even in the best-case scenario where the null hypothesis can be rejected, statisticians remind clinicians that the treatment effect is estimated with uncertainty. This makes them even more unfriendly. Let's face it: Statistics are very frustrating to clinicians, whose job it is to use the new treatment or not, rather than to procrastinate about errors the way statisticians do. This cultural difference does create friction in the collaboration between clinicians and statisticians; more importantly, it also causes clinicians to look for quick fixes that typically don't work, such as using non-randomized evidence (at the risk of biases) or conducting small clinical trials (at the risk of large random errors).

But friction can also be positive for both parties: Clinicians have improved the way they conduct clinical trials because of the scrutiny of biostatisticians; conversely, statisticians learn much from clinicians, who push them in the right direction to develop a statistical methodology that meets their requirements.

13.2.5 The Statistician's Oath

Now taking a broader view of 'the truth,' I believe it would be suitable for statisticians to adopt an oath similar in spirit to that used in sworn testimony. For example:

> *I shall analyze data*
> *to uncover the truth,*
> *the whole truth, and*
> *nothing but the truth*

Here, 'the truth' might refer not just to the true value of a parameter, but it might also encompass a complete description of the experimental conditions, assumptions made in the analysis, etc. Such an oath would then adequately cover most situations in applied statistics and would strongly suggest to include, in the reports of statistical analyses, a level of detail that is often ignored or brushed off as inconsequential in the interpretation of the results. Consider, as an example, the results of a randomized clinical trial that are discussed for a large population of patients when such results do not generalize from the sample of patients entered in the trial. This is clearly in violation of the *nothing but the truth* part of the oath, but it is certainly easier than suggesting another trial is needed to confirm the results in the larger population. Consider, as another example, the results of a causal analysis without mention of the assumptions made. This is clearly in violation of the *whole truth* part of the oath, but it is far easier than to explicitly list assumptions that are not verifiable at best, or untenable at worst.

I am not sure having a professional oath for statisticians would be feasible, but the point is to make the purpose of the statistical profession obvious to all. In our troubled times, where lying is equated to having a different opinion, a formal oath would be a welcome confirmation of our main professional obligation. Note that such is not the case of the legal profession, for lawyers must often defend their client at the expense of, rather in the service of, the truth. Sworn witnesses must tell the truth, but the lawyers are under no such obligation, and for good reason. The distinction implies no moral judgment, but it must be made clearly. I have often represented pharmaceutical companies at meetings with the Food and Drug Administration or European Medicines Agency. In these meetings, I have always acted as a scientist and never as a lawyer: I would always speak my mind, and I would never take a position that I knew to be incorrect just to defend the interests of a client.

13.2.6 On Cheating

It is a capital mistake to theorize before one has data.

Insensibly one begins to twist facts to suit theories, instead of theories to suit facts.

ARTHUR CONAN DOYLE (1859–1930),
A SCANDAL IN BOHEMIA

Cheating is the active form of lying. In a sense, cheating is lying to oneself, since one deliberately acts in a manner that one knows not to be correct or desirable. The honest investigator tries to find out whether a theory is correct by seeing how well the theory fits the data (just as Galileo did). The dishonest investigator, in contrast, tries to convince others that a theory is correct by fitting the data to the theory.

I've always been fascinated by cheating: Why do people cheat? My interest, by the way, is about cheating in science, for instance by manipulating data, not cheating on one's spouse or companion (a problem far too complex to be discussed here). An interesting theory posits that cheating is part of our self-defense heritage. Animals cheat to escape their predators: they change form, color, smell, or anything else if it protects them from immediate dangers.

The ability to cheat is a great competitive advantage, and it comes as no surprise that human beings have acquired more than their fair share of this inherited trait. Works of fiction, which human beings enjoy and value so much, are the imaginary equivalent of cheating: much satisfaction is derived from seeing others cheat and get away with it. Cheating is not just biologically advantageous and socially inevitable; it is also emotionally satisfying – if undiscovered!

I became interested in cheating when I heard about a large clinical trial (7,000 patients) in which one investigator had entered more than 400 patients, none of whom he had treated with the treatments under investigation. This investigator had taken data from a historical database of similar patients and had simply copied these data onto the case report forms of the trial. It seemed to me that in order to make these data plausible, the investigator had to tweak them a little, which presumably created systematic differences between his data and the data from all other investigators in the trial. Such differences were therefore potentially detectable using statistical methods.

My colleagues and I developed such statistical methods (based, essentially, on mixed effects models). We later founded a company to make these methods available to clinical trial sponsors (Venet et al., 2012). Our methods did detect the sorts of fraud that had initially triggered our work, but also, more interestingly, other serious data issues that typically remain undiscovered by trial sponsors. The power of using statistics to detect irregularities in data is yet to be applied to many other situations.

13.2.7 On Collaborating

Great things in business are never done by one person; they're done by a team of people.

STEVE JOBS (1955–2011)

The collaboration on statistical methods for fraud detection was one of the most enjoyable of my career, but there have been many others. This may well be the most remarkable change in science over the last decades: The Internet has made science collaborative in nature, experiments can be conducted by thousands of investigators simultaneously, and results are shared in real time with the scientific community worldwide. In 1993, the Ig Nobel prize in literature was awarded to the 977 authors of a medical research paper that had 100 times as many authors as pages (The GUSTO Investigators, 1993). A decade later, a physics paper set a record with more than 5,000 authors; more pages were needed to fit the authors' names than the results of their work (Castelvecchi, 2015).

Although these examples seem hilarious, they do reflect the reality of science today, and exemplify the amazing power of 'distributed science,' which – just like distributed computing – is manyfold more powerful than the isolated discoveries of the past. What's better is that at a personal level, it is far more gratifying as well as productive to work on a problem as part of a team. Even more so in an international context, where cultural differences make the collaboration richer and more fruitful.

The greatest joys of my professional life have come from working with like-minded individuals toward objectives that we all profoundly believed in. I feel very privileged that many of these collaborators have become friends whom I continue to see and interact with, even when our research interests have diverged. Figure 13.3 is a photo taken in the beautiful mountains of Hokkaido (Japan) at the end of a collaboration aimed at estimating the effects of various treatments of gastric cancer between statisticians from Japan, Belgium, and France. The smiles on the picture reflect our shared joy for a collaborative work well done. Pictured from left to right: Koji Oba, Valerie and Marc Buyse, Tomasz Burzykowski, Yasuo Ohashi, Harry Bleiberg, Xavier Paoletti, and Junichi Sakamoto)

FIGURE 13.3   Members of The GASTRIC collaboration take a break from their work to enjoy the mountain scenery near Hokkaido, Japan

13.2.8 On Getting Papers Published ... or Not

Human institutions are so imperfect by their nature that in order to destroy them, it is almost always enough to extend their underlying ideas to the extreme.

ALEXIS DE TOCQUEVILLE (1805–1859)

Another joy that comes with collaborations is the opportunity of writing papers for publication in statistical journals, or contributing to papers for publication in medical journals. At least I find it a joy, though most consider it a pain. Where it does become a pain is when the paper is ready for submission. In the old days, submission required to have five printed copies of the manuscript, to put them in an envelope and send them to the journal, and voilà! After a few months of patience, the reviewers' comments would come back by return mail, and the process would be repeated as many times as needed (typically once).

In our days of electronic speed, the submission itself tends to take more time than the writing of the manuscript, edits and revisions included. The reviewers' comments, though sent electronically, come back no faster than ever before, and single peer reviews have become a rare occasion. I still enjoy writing papers, but I have lost interest in submitting them to any other journal than arXiv or medRxiv, where submission merely consists of uploading the paper without peer review, and making it open access without any restrictions thereafter. My bet and hope is that the arXiv/medRxiv model will eventually prevail, with official journals publishing only expert commentaries on the papers available in open-access journals.

Will papers lose much of their credibility if they are not peer-reviewed? I think not, based on the early days of the COVID pandemic, when many excellent papers appeared instantaneously in arXiv, while many awful papers appeared after peer review in supposedly more reputable journals (remember the peer-reviewed publications about hydroxychloroquine?). Let's face it:

The peer-review system is inefficient and unsustainable at best, or counterproductive at worst. It started with a wonderful intention, but it ended up in absolute misery. Reviewers are expected to do a quick and thorough review, without payment or recognition, and without any bias or prejudice. That's a lot to ask for, probably too much to ask for in most cases.

Of all the limitations of peer review, the most damaging may well be the unavoidable biases of experts who work on closely related topics … even for statisticians whose job it is to eliminate all sources of bias, even from their own opinions. It is a testimony to the honesty and dedication of researchers that so many of them accept to contribute to a broken system. Who doesn't have stories to tell about reviewers being completely wrong or misguided, insulting in rare cases but too often arrogant or condescending?

On the lighter side, I once received a comment from a reviewer who found it useful to copy the conclusion of my paper, but with a typo in it: 'It is likely that profession-free survival is a good surrogate for overall survival.' Steve George told me of a reviewer who had objected to his use of the 'impaired t-test.' Typos and jokes aside, the peer-review process cannot be described as anything else than an attempt to have a thoughtful piece of work destroyed by individuals lacking time, motivation, impartiality, or competence to do a good job at it. Now is the time to move to open-access, non-peer-reviewed literature.

13.2.9 On Failures

Success is walking from failure to failure with no loss of enthusiasm.

WINSTON CHURCHILL (1874–1965)

I had been running my own company in Belgium for a few years when I was contacted by a prestigious University on the East Coast of the United States where a position had just become vacant in the biostatistics department. My background and experience were apparently a great fit for the position they were looking to fill, so

I was encouraged to compete to get the position. I participated in the first round of interviews, and did well. I was becoming quite excited at the prospect of leaving my position as self-appointed CEO of a small company to join academia. As luck would have it, an excellent colleague who had left my company a few years earlier contacted me and indicated his interest in coming back. I jumped on the opportunity and offered him my job, as I was certain to leave soon. I participated in the second round of interviews, and felt I had again done quite well. A few days later, I was officially informed that the University had decided to appoint a better-qualified candidate. I sent emails thanking all the individuals who had supported me, and remained in limbo for a whole day, crushed and devastated that such a wondrous shift in my career was not going to happen. And indeed, it never did.

Instead of moving to the highly competitive environment of a campus on the US East Coast, I asked my wife if she would like to move to the West Coast, where I would work remotely for my company, while spending time working on overdue research projects. She quickly approved of the plan, and a few months later we were on our way to San Francisco.

Years later, as I reflect on this episode which I took as a monumental failure, I realize that it turned out to be, in fact, a surprising opportunity. I am aware that I look at these events today with a great deal of posthoc rationalization, but I find it hard to reason counterfactually and assess what my life would have looked like, had I gotten this position. In the life that I actually lived, my successor got firmly installed in the company, while I moved to sunnier skies, got used to spending days in T-shirts and shorts, took up biking on the hills of the Bay Area, and carved out more time to do research than ever before. No longer did I have to take care of management issues, and I could devote time to eccentric activities such as learning to fly a plane.

Stanford psychologist Laura Carstensen and her colleagues have conducted studies to show (and explain why) people get happier as they get older (Carstensen et al., 2000). My personal experience is congruent with her findings, though I am ready to accept that in my case, moving to California may have been a strong confounding factor.

13.2.10 On Taking Risks

Your mistakes might as well be your own, instead of someone else's.

BILLY WILDER (1906–2002), FILMMAKER

Statisticians love to control risks – that's what they are trained for. But they hate *taking* risks. The American Statistical Association held its Joint Statistical Meetings (JSM) in Las Vegas from August 5, 1985 to August 8, 1985. The story goes that the casino revenues dropped to an all-time low during that week. I do not know if the story is true, but it might very well be: Statisticians do not play games of chance, because they know they'll lose on average.

However, life goes beyond the average: Deviations and outlying values are worthy of consideration, too! It is interesting to observe that we currently tend to appreciate extreme observations and chaotic behaviors, while earlier periods focused primarily on averages and predictable behaviors. In my own country, Adolphe Quetelet is remembered as the astronomer who founded the Observatory of Brussels and as a prolific and influential statistician and sociologist (Figure 13.4). Quetelet was obsessed with applying the astronomers' 'law of errors,' i.e., the normal distribution, to human features such as height and weight (he introduced the body mass index which is still in use today) and many other human traits. He developed the concept of the 'average man' who would have the average value of all the measurable human traits. Porter (1986) argued that the average man concept accorded well with the ideals of the bourgeoisie, who saw deviations from the mean as imperfections and was generally suspicious of excesses in politics as well as in social science.

Today, for better or worse, it is the excesses that make the news, while the average is considered dull and uninteresting. Such is

FIGURE 13.4 A Belgian stamp of Adolphe Quetelet (1796–1874). The Belgian branch of the International Biometric Society (IBS), established in 1952, was named the Adolphe Quetelet Society.

the pendulum of history, but the statistical professions may still be too attached to the mean to consider opportunities associated with extreme, less frequently observed, values.

In my early forties, I decided to invest a modest (though substantial relative to my income) amount of money in creating a business. Many of my statistical colleagues and friends were skeptical: The vast majority of them thought I was crazy to leave an interesting, reasonably well-paid job to be exposed to professional hazards and potential hardship. And yet this jump into the unknown led to many surprises, good and bad, none of which I would want to avoid even with the benefit of hindsight.

13.2.11 On Successes

Experience is a comb that one receives when one has gone bald.

BERNARD BLIER (1916–1989), FRENCH ACTOR

I find it odd that I remember my failures much better than my successes. Is it because I tend to remember failures and forget successes (recall bias), because I label 'failures' differently than 'successes' (observation bias), or because there were truly more failures than successes (a hypothesis worth considering – much like in clinical research, where more trials fail than succeed)? I don't know, and it does not matter, but the point is that we are educated to achieve success, not to handle failure. We do not feel nearly as much joy in succeeding as we feel pain at failing – and this negative bias is built into us in our early years.

Education is in acute and urgent need of reform. Current educational systems tend to be oppressive and ineffective. They often favor abstract thinking over emotional intelligence, reward obedient memorization, and limit freedom to explore and imagine. We spend our lives trying to adapt to a world that education has not prepared us to confront. Some can take it, but many cannot, and we end up with a society built on dissatisfaction, fear, negative feelings, and violence.

I realized all this during a very painful episode my girlfriend and I went through (yes, another failure of mine). We had decided to go on our own separate ways, and she had sent me a small book entitled *The Four Agreements: A Practical Guide to Personal Freedom* (Don Miguel Ruiz, 1997). I usually do not care much for how-to, feel-good books. It took me a long time to begin reading this one, and I had to be reminded of its presence on my nightstand several times before I started reading it. This little book changed my life (beyond the fact that my girlfriend and I came back together, and later got married). If you have not read this book, there is a good chance it might change your life, too. At $7.74 in paperback, it is a great investment in yourself.

CHAPTER 14

Perspectives on Careers

Geert Molenberghs

ABSTRACT

Starting from his earlier prior education, Geert Molenberghs sketches the beginnings of his academic statistical career. Geert then discusses research and creativity, and how collaboration in its many forms can benefit that work. Geert's career journey illustrates how interactions with colleagues in settings somewhat different from our day-to-day work can benefit creativity. He shows us the crucial role mentoring and the learned societies play in academic life and in advancing science. Geert's journey illustrates the diversity in skills required, many of which need to be learned on the job, over a career in statistics. The research, teaching, and consultancy required of many academic positions are already a diverse set of responsibilities, but many academic statisticians also take up roles in leadership and administration, which require still different skills. Geert's unique role in the

DOI: 10.1201/9781003334286-19

COVID pandemic illustrates the challenges in communication to non-scientific audiences and how we can meet those challenges. Most of all Geert's career is a good illustration of work–life balance as he shows us the role hobbies and sports activities have played in his life.

14.1 INTRODUCTION

Reading this volume and preparing for my own contribution forced me to look back on the past 30 years of my career, and nearly double that time when considering my entire life. In retrospect, it is the events that happened which fill the book of one's life, but when I started, I could not have imagined the course of my life: A career based on academic training yes, a career in academia not really, let alone in statistics rather than in pure mathematics. And the last thing that I would have imagined is that one day people would recognize me on the street and in the supermarket because of my public role during the COVID-19 pandemic.

Like so often, the script of my life got rewritten several times while the movie was being filmed.

14.2 THE BEGINNING

I was born in the city of Antwerp, Belgium, in a working-class family. I had the good fortune to grow up in a loving and caring family. Thanks to a good school system, the fact that my parents chose a school for me that emphasized mathematics – many students went on to engineering, mathematics, and related fields, – that I really liked mathematics, and that I did well in the subject, I naturally was set up to go on to higher education in mathematics. So, I started what is now called bachelor's and master's in mathematics at the University of Antwerp – four years of study. In those days, I really enjoyed the very abstract subjects best, such as calculus, geometry, and algebra. When the time came to select a specialty, in the third year, it was almost self-evident to pick pure mathematics; we were only three 'pure students' in a group of about 20. When I graduated in 1988, I secured a PhD grant from the Belgian Science Foundation, so I embarked on a PhD in algebra. However, in hindsight, I did not have the enthusiasm required for the subject in those early days of my PhD research as I had felt during my bachelor's and master's years.

I belonged to one of the last contingents that had to fulfill compulsory military service. One could opt for a substitute 'civil service,' e.g., at a government administration, a non-governmental organization, or, like in my case, in a healthcare facility. My PhD advisors were aware that at the Academic Hospital in Leuven, Belgium, they were looking for someone planning to do civil service, so I went to work for 20 months for Emmanuel Lesaffre. To my good fortune, Emmanuel did not give me routine jobs, but involved me in the research of the department. I programmed likelihoods – in GAUSS – and optimization routines, applied them to real data, and to my amazement, meaningful results came out. Biostatistics was love at first sight! I had had some courses in measure theory, probability, and statistics, but nowhere near the hands-on work that I could do here.

Because of my Science Foundation research grant, I had some flexibility and, after jumping through some hoops and over some hurdles, I was able to carry on research in statistics rather than algebra. So, when my 20 months of civil service were over, I simply continued working on marginal models for multivariate categorical data. At one point, when Mike Kenward was visiting, Mike, Emmanuel, and I realized that my work was of value for repeated measures as well, and because of its likelihood basis had advantages when data were incomplete.

Lesson learned: There is no reason to worry or despair when your trajectory through education and early research is not textbook – in fact it rarely is.

On January 29, 1993, I defended my PhD; two days later was the deadline for an assistant professorship at Hasselt University.

I was bold enough to apply and, to my amazement, I got the position! I started on May 1, 1993, but my colleagues at Hasselt University, realizing that I was about to skip the typical postdoctoral phase and lacked experience, sent me to the Harvard School of Public Health, where I stayed for the rest of the year. This was an eye-opener! I had been to conferences, on research visits to the United Kingdom, and during my final master's year, I had lived in Paris for three months (I have loved Paris ever since), but spending half a year in a place where there were more statisticians than in my entire country was something else.

Hasselt University was one of the first universities to offer master-level training in biostatistics in continental Europe. To staff the program, they considerably relied on visiting faculty, especially in the early years, with many coming from Harvard. In 1993, Louise Ryan and Stuart Lipsitz were on the visiting faculty, and the late Steve Lagakos visited too. Steve hosted my visit at Harvard, and I was able to do work with Louise and Stuart. Academic life, the way it was organized in North America, never left me, and I have to say that our institute is organized pretty much in a similar way.

The sheer size of statistics at Harvard, as well as in so many North American institutions, and the realization that in most European countries the scale was much more modest, and a lot of building was to be done, made me reflect on what to do: go back to Europe or stay. I decided to go back.

> Lesson learned: Some people function best in a large, well-oiled organization, while others thrive when they can roll up their sleeves and start building.

14.3 PRODUCTIVITY, PRIORITIZATION, AND COLLABORATION

Evaluating my Harvard visit with Steve Lagakos at the very end, I was happy enough that he thought it successful, but he warned me that I ran the risk of getting involved with too many things at the same time. He also warned me that about everything in our academic life has hard deadlines, except … research, and that a simple trick is to assign to a paper, a book chapter, a book, or any other research project a deadline, and then have the discipline to stick to it. It works better in projects with coauthors, if everyone is on the same page!

> Lessons learned: Prioritization, focus, and organization all play key roles in consistently getting important work done on time.

I have always enjoyed doing research and considered it a hobby more than work – which is not to say that it is not taken seriously, quite on the contrary. We all know how seriously many people take their hobbies, and so do I, just as research. The great thing about biostatistics, as well as many other forms of applied statistics, is that it combines mathematics and one or more substantive sciences. These diverse collaborations can be fun, enlightening, and can lead to valuable contributions to health, public health, and food production.

Most academics, and I am not an exception, are also expected to teach and engage in consultancy. Consultancy was not so common when I joined Hasselt University and part of my assignment was to build a consultancy unit there. I have always considered both teaching and consultancy fun parts of the job, but just a bit more 'job' than 'hobby.' That is perfectly fine, each colleague has their own preference and the time available is divided differently over research, teaching, and consultancy for different people. However, we have always considered these three pillars equally important and equally valuable for the institute as a whole.

Furthermore, the three pillars have a high potential for virtuous interaction. Many successful statisticians around the globe turn consulting problems into a highly relevant, statistically and substantively, research agenda. In 1994, we provided consultancy to the then-starting Belgian Health Interview Survey. Over the years, this has led to many papers, and I have been using the case in various sampling theories and other general courses over the last two decades.

The work that I have been doing over the past decades with Geert Verbeke has led to various papers and, in particular, two books on longitudinal and incomplete data (Verbeke & Molenberghs, 2000, Molenberghs & Verbeke, 2005). Of course, these books include collaborative work done with a large array of fine colleagues. The books, products of scientific work, have had ramifications for teaching, consultancy, and administration and organization.

Let me try and explain that. First, we have taught about a hundred short courses on the topic, as well as regular courses in our home universities, UHasselt and KU Leuven. Second, in the slipstream of these courses, it is not uncommon for short course instructors to be approached by course participants with the request to further embark on a consulting relationship. The administration for such consulting has to pass via the university, but in our case, that was plural: Universities. Contract negotiations can be involved enough, because of legal, financial, and intellectual property issues, but having to do this with two universities was even more cumbersome. Third, and related to this, we concluded that our two biostatistics groups really had a long tradition of collaboration, extending well beyond ourselves, through various research lines, and collaboration between the Master of Statistics programs of the two universities. One thing led to another, and in 2007 the Interuniversity Institute of Biostatistics and Statistical Bioinformatics (I-BioStat) was founded. It really was a bottom-up initiative, for which at first the academic authorities of both universities had to be won over. We decided not to make the structure too complex, but rather settle for subsidiarity: Whatever could more effectively be done at one of the two entities was kept that way, otherwise we worked together. This avoided long and tedious 'merger' discussions and implies that staff members are properly integrated within their own university, which is politically and diplomatically important.

> Lesson learned: The pairwise and three-way interactions between research, teaching, and consulting are critically important for the healthy life and growth of a biostatistics group.
>
> Lesson learned: When there is a true spirit of collaboration and trust, one and one truly make more than two.

Collaboration is by no means restricted to colleagues either within academia, within industry, or within the government. I have had the pleasure of being involved in 'between' collaboration. Facilitated by Ray Carroll, Craig Mallinckrodt and I started joint work on missing data issues, gradually involving a number of other topics and various colleagues. Over the years, this has led to several papers; one that I am particularly proud of appeared about two decades ago, and involved, apart from Craig, also Ray Carroll and Mike Kenward (Molenberghs et al., 2004).

I had the pleasure and honor to teach various short courses at Joint Statistical Meetings with Alex Dmitrienko and Christy Chuang-Stein on analysis methodology for clinical trial data. Here too, one thing led to another and, apart from various manuscripts, we jointly authored a book in the SAS Press Series on this topic (Dmitrienko et al., 2005).

14.4 CREATIVITY AND INNOVATION

Creativity takes many forms and arguably depends on the field one is working in, the personality of the researcher, etc. Sometimes, one merely needs to be creative toward solving a problem already posed, such as one presenting itself through consultancy. On other occasions, defining an interesting research problem can be a creative act on its own. For me, and I guess for many people that does not happen by simply blocking off time for it in my calendar. Literally, interesting problems, and nice solutions to problems already defined, may pop up anywhere, at any time. They may present themselves while cooking, shopping, running, waiting for the train, or while taking gas.

14.5 COLLABORATIVE RESEARCH, AND LEARNING FROM OUR PEERS

Problems and solutions are not always a solitary act, at least not for me. A conversation with a trusted colleague, sometimes even a random conversation, one that is going all over the place, may suddenly uncover a meaningful lead. What do I mean by a trusted colleague? When solving a problem, it is perfectly OK to follow leads into dead-end streets, perhaps over and over again, until suddenly progress is made. The prerequisite is that one should allow oneself and one another to make mistakes. It is more productive to loosely brainstorm than to try and outwit each other.

In April 1994, I had a conversation with Marc Buyse at a DIA meeting. He asked me a question about the calculation of a confidence interval for the so-called relative effect, a quantity now well known in the context of surrogate endpoint evaluation in clinical trials. My answer was, 'Wait a minute …' Rather than diving into the technicalities, I wanted to think a little bit about the advantages and disadvantages of the quantity the confidence interval was intended for. The rest is history, we have been working together for three decades on the topic, with a large group of people from all over the world on the topic, leading to a good number of papers and two books (Burzykowski et al., 2005; Alonso et al., 2017).

This not only happens between statisticians. Arguably, it is even more prevalent in conversations between statisticians and substantive scientists. A classical example is when a targeted question is asked about a particular method (e.g., which multiple comparisons correction to use with ANOVA), whereas the method itself is not appropriate, or at least not optimal, in the first place (e.g., data are hierarchical and incomplete, and a mixed model is more advisable).

> Lesson learned: A useful skill for a statistician is not to answer a question when asked, but to think about the question first and perhaps reformulate it.

When working with peers, we should do our very best, without assuming that we are the best. Rather we may have something to contribute in one or a few aspects, while we can benefit from the skills and talents of others in other areas. A collaborative research project benefits from interaction between people who are creative, have ideas, and see opportunities, and others who see the roadblocks on the way. When only the first is present, a project might enthusiastically crash into a wall. With only the second tactic, the project might never get started. Someone with extraordinary research skills may actually be a poor communicator, where other people may be masterful teachers, able to explain a method, a result, etc., much more clearly than the people who did the work. That's not a problem, quite on the contrary – problems start when we fail to acknowledge this.

> Lesson learned: It is a wise strategy for applied statisticians to seek for the optimal combinations of all team members' skills and strengths, rather than hoping in vain to be the best in everything oneself.

We do not only work with fellow statisticians, but often involve in interdisciplinary and multidisciplinary work. It is not uncommon for people in our profession to be the only statistician in the boardroom meeting. A typical example is a Data Monitoring Committee (DMC) meeting for a clinical trial. In an open meeting, there may be a few statisticians among a much larger clinical group. In a closed meeting, one statistician may be interacting with a group of clinicians. A lot has been written about it and the book by Ellenberg et al. (2019) is wonderful in this regard, but one aspect I would like to highlight is that often we as statisticians see issues that colleagues from other professions do not see – and of course vice versa. But this implies that we may literally be the only ones in the room to think of the fact that, for example, 2:1 randomization tends to be less precise than 1:1 randomization and that it is much more difficult to keep the blind for DMC members. At such a point, we will not be asked for our opinion on the issue, as others may not see an issue at all.

We therefore need to speak when we are not asked to – it is much better to put a non-issue on the table than to remain silent about an issue. Of course, we should be flexible enough to understand that the final decision will be a compromise, where statistical considerations are but one component. For example, our clinical colleagues may argue that for them it is easier to recruit patients if they have a 67% chance to receive the new, promising medication.

> Lesson learned: Speak when you are not asked to.
> Learned Societies, Scientific Journals, and the Evolution
> of our Profession

When working in an academic field such as statistics and biostatistics, we should realize that it is not a perpetuum mobile – it requires energy and labor to keep it going. Whether we work in academia, industry, or the government, we have to acknowledge that our affiliation is not only with the entity that produces our paycheck but also with the professional and/or academic field to which we belong. For applied scientists, affiliation and identification might be with fields – plural – rather than field. I tend to identify myself with statistics, biostatistics, epidemiology, public health, clinical trials, data science, statistics in humanities and social sciences, etc.

Of course, 'profession' is rather vague and etheric. But a very tangible way to express our professional affiliation is through membership in learned and professional societies and organizations. Stimulated by Emmanuel Lesaffre from when I was a PhD student, I have always considered it self-evident, not only to be a member of such societies, but gradually also to take up various roles. Table 14.1 provides a selection of such roles. Roles not mentioned in the table involve helping with the organization of scientific meetings (scientific and organizing committees; session organizer; short course instructor; panel member or moderator, etc.).

My IBS presidency came, once more, as a coincidence. I was serving the IBS as General Secretary from 2001 onwards. In 2003, Rob Kempton became the incoming Vice-President. Following IBS procedures, he would have become President in 2004 and

TABLE 14.1 Membership and Roles in Learned Societies

Society or Organization	Current Roles	Past Roles
Royal Statistical Society of Belgium	Vice-president	Treasurer
International Biometric Society	Executive Editor Biometrics; member of Editorial Advisory Committee	President (2004–2005); Vice-President (2003 and 2006); Biometric Bulletin Editor; General Secretary (2001–2003); Council Member
International Statistical Institute		Life Sciences Committee Chair
Royal Statistical Society (UK)		President Nominating Committee, Editorial Policy Board, Programme Committee, International Committee
American Statistical Association (USA)		Member of the International Relations in Statistics Committee (chair 2014–2017)
Quetelet Society (Belgian Region of International Biometric Society)		Treasurer

2005, were it not for his sudden and untimely death in 2003. This had never happened before in the society, and an ad hoc election was organized late in 2003. The person elected would enter the vice-presidency for barely a couple of months, which would then turn into the presidency. Since I was familiar with the executive committee from my secretarial role, I decided to run and was elected.

Over time, I realized that scientific societies gradually began to be called into question, in the sense that younger colleagues were no longer clear on the benefits of and hence the need for society membership. As I see it, there are two main reasons for this. First, publication became electronically available, rather than 'print only,' so the most tangible benefit of being a society member – receiving the journals – became less of a driver because the electronic version was often available through the library's consortia deals.

Second, and perhaps more fundamental, propelled by a number of scientific revolutions, people considered themselves less and less affiliated with a single field. This was very clear at the time of the statistical genetics revolution (turn of the millennium), the advent of big data, the coming of data science, etc. I devoted a large part of my so-called presidential address to it when I was president of the International Biometric Society (Molenberghs, 2005). I am of the opinion that these so-called revolutions are actually evolutions – rapids in the profession's river. When the dust settles, it is still necessary to organize scientific meetings, ensure high-quality peer review and scientific publications, and have advocacy groups for the profession.

It is true that affiliation with more than one area is the norm rather than the exception, and learned societies should acknowledge that, broaden their scope, and reach out to each other.

Lesson learned: Paraphrasing John F. Kennedy's words in his inaugural speech: 'Ask not what your profession can do for you – ask what you can do for your profession.'

A key player among professional organizations is the academies of science and medicine. I have been a member for a decade of the Royal Academy of Medicine in Belgium, among clinicians, pharmacists, nursing scientists, veterinarians, and other health professionals. In Belgium, like in many countries, the academies play a role in formulating opinions about key (public health) questions. Generally, they form a stable base of impartial wisdom.

A key theme is scientific publication. I have been fortunate enough to experience first-hand all sides of it. Apart from having been an author and referee at multiple occasions, I have also served in Associate Editor, Editor, and Executive Editor roles. An overview of the latter is given in Table 14.2.

TABLE 14.2   Peer Review, Editorial, and Related Roles

Period	Society or Organization (Constituency)	Role
1998–2001	International Biometric Society	Editor of Biometric Bulletin (Newsletter)
1998–2001	Royal Statistical Society (UK)	Research Section Committee, Member
2001–2004	Royal Statistical Society (UK)	Joint Editor of Applied Statistics
2003–2018	Wiley	Member of the Wiley Series in Probability and Statistics Editors team
2007–2009	International Biometric Society	Co-editor of Biometrics
2010–2015	Oxford University Press	Co-editor of Biostatistics
2013–2018	Wiley	Member of the Wiley StatsRef team
2018–	International Biometric Society	Executive Editor of Biometrics
Various Associate Editorships		Biometrics, Applied Statistics, Archives of Public Health (Belgium), Biostatistics, Canadian Journal of Statistics

Many journals in our field use a hierarchy of peer-review roles, where an editor screens a paper, and then either rejects it or sends it to an Associate Editor. The Associate Editor may report back without further ado, or send the paper on, in turn, to one or more referees. All of these require a substantial amount of work, and it is not always clear that this work is valued by our organizations. Nevertheless, it is a crucial component of the scientific fabric. Without peer-reviewed scientific communications, progress would grind to a halt. So, we depend on the goodwill of our colleagues to take up such roles. From the postdoctoral stage onwards, being a referee provides valuable experience and trains critical thinking. Many junior colleagues – and of course also senior ones – make very good Associate Editors.

As is clear from Table 14.2, I had the pleasure to be journal editor for three journals. Two of those, *Applied Statistics* (a.k.a. *Journal of the Royal Statistical Society*, Series C) and *Biometrics* are society-owned journals; the former by the Royal Statistical Society, the latter by the International Biometric Society (IBS). *Biostatistics* is owned by Oxford University Press, a commercial publisher, but with very strong ties to the academic community. Scientific publication has been going through major changes, with first the move to electronic publication and then to Open Access. Along the way, large consortia deals saw the light of day.

This evolution is not finished. In my opinion, the involvement of learned societies is crucial. Leaving academic publishing solely to the market implies that scholars are doing considerable work on a voluntary, non-paid basis, while profits are reaped by the market. Even more importantly, the free market may not offer the best guarantees for academic freedom. The quality of our work should be judged by our peers – and our peers only. This implies commitment, and work, from all of us. This statement is not anti-market operation, but rather suggests a good balance between market and professional communities.

Thus, independence and quality are keywords – much more than quantity. Unfortunately, because of academic evaluation systems in vigor in many places, quantity tends to get too important a role. Publishing becomes a goal of its own, preferably in journals with as high an impact factor as possible. Sometimes bitter fights over author positions tend to follow in the wake of this. Arguably though, when writing a good paper, it helps to keep the audience in mind. One should be convinced that there is value for this particular receiver to receive this particular message, and one should better make certain that the message is clear, unambiguous, and communicated in an engaging way. The best choice for a journal may not be the one with the highest impact factor in my subfield, but rather the one that is closest to the target audience.

> Lesson learned: A paper or other scientific publication is a message, intended to go from a sender to a receiver. Merely reducing it to an item on one's CV is unhealthy. Lesson learned: We cannot expect that our papers are professionally and timely peer-reviewed, whilst refusing all peer-review work ourselves.

Peer review is not restricted to journal manuscripts, but covers grant applications, reviews of departments in universities and academic institutes, etc. A dilemma is that people who tend to do a good job are solicited time and again until … they are no longer able to do a good job. One needs to find a good balance between the various tasks for which our input is solicited, and that implies that not every request can be honored.

> Lesson learned: Even good citizens sometimes have to say 'no.'

In the above examples, it will have become clear that my affiliation with learned societies is not limited to Belgian ones, even though I was born there and that this is where I have been working. Belgium is a small country, with about 11.5 million inhabitants.

One should do one and the other: Bond with national and international colleagues alike. In a large country with a broad, time-honored, and well-established statistical infrastructure, such as the United Kingdom and the United States, it is tempting to stay within national limits. I would argue though that even then affiliation with international societies and other constituencies, including those building bridges between the Global South and Global North, is a proper civic action from which we strongly benefit.

14.6 LEADERSHIP AND ADMINISTRATION

Taking up leadership and administrative roles, in one's own institution as well as in other settings, such as learned societies, may at one time be unavoidable. That is not meant as a negative or fatalistic statement.

It can be advantageous for statisticians to take up higher-level administrative and leadership roles but of course, time spent on administration is time not spent on research, education, and consulting. A team of statisticians should therefore regularly strategize on the level of its involvement in administration, and who would be best suitable for it. Use people's talents wisely. Promising, bright, young faculty members might be ideally suited for it, only ... once they are deeply involved in such roles, their real, and perhaps rare academic talent may be lost. There is a time for everything ...

A worldwide phenomenon is that central supporting services in organizations, whatever their nature, tend to keep growing at perhaps too fast a pace, much faster than the productive, academic departments. For the health and benefit of our organizations, we should not stop hammering the nail that administrations should be lean and mean, and service-oriented.

All too often, a successful leader is convinced of their own unique talent, status, and position. This applies to countries and academic departments alike. Once one starts to think that one is the uniquely best-positioned person to fulfill a certain leadership role, it is best to make way for someone else. When I was appointed at Hasselt University, the statistics group was relatively small. There were three statistics professors who took all decisions together. Ever since we have been operating under a model of shared leadership. For example, in our interuniversity institute I-BioStat, the overarching director works together with the entity-specific directors so that in actual fact most decisions are taken jointly. This averages out idiosyncrasies and helps maintain balance. Many learned societies have an executive committee, thus avoiding that a single person can take all decisions.

Of course, when a crisis strikes, swift action may be called for, and then the command lines should be clear – when the ship is heading for the iceberg, it is not the time to set up a focus group to decide whether the steering wheel should be turned, and if so by whom. But in all other settings, regular consultation is darn useful.

One of the common 'founder faults' is that the founder is attributed such a unique status that no serious thought is given to succession. All of us know examples from history. Good founders should ensure that they are surrounded by people equally good or better than themselves, and then set up their organization such that even when they would suddenly disappear, the organization would keep running smoothly.

Somewhat in the same spirit, I have always believed that giving autonomy to colleagues, including junior colleagues and PhD students, is the shortest way to excellence. This is especially true for scientific research, an act of creativity. We tend to stimulate people to interact, so that senior, mid-career, and junior colleagues can all bring in their specific skills and learn from each other. Arguably, mentoring is more fruitful than authority – and when there is going to be an authority, let it be a natural authority. While many colleagues with academic training thrive in such an environment, for some the lack of a strictly imposed structure may be stressful, the rebound of which may be that they tend to establish a mini-hierarchy around them – something to be watchful for.

A fine part of academic life is the ability to work with PhD students. Over the years, I have found that working in teams, with various PhD students and/or several senior colleagues, is very effective. Everybody brings in their specific expertise, ideas, and perhaps also preferences; this will make for a rich discussion. Depending on personality and cultural heritage, some junior colleagues find it difficult or even 'not done' to correct mistakes made by more senior colleagues. However, I have always taken the attitude that, mathematically, 1 and 1 is 2, and the most senior colleague who claims otherwise makes a mistake. We are all served well if an error (e.g., in a proof, a model formulation) is corrected, no matter who made the error and regardless of who corrects it.

> Lesson learned: Contentons-nous de faire réfléchir. N'essayons pas de convaincre. (George Bracque)
>
> Lesson learned: Administration is needed, but with moderation.
>
> Lesson learned: Nobody is irreplaceable, especially not the leadership of an entity.

14.7 COMMUNICATION, IN NORMAL TIMES AND DURING THE COVID-19 PANDEMIC

Most of us are involved in communication on a routine basis. We write papers, give talks at conferences as well as seminars, teach regular and short courses, and explain statistical matters to our substantive colleagues. All these take place in a protected, academic environment, far away from the spotlights.

Occasionally, we communicate to wider audiences. For example, we may communicate through regional or national media about a clinical (trial) finding or a result of an epidemiological study, sometimes alongside a substantive colleague. This itself is a challenge. We are trained to be accurate and complete, with an eye for (technical) detail. Well, that is not going to work on national news, where the message must be folded into a two-minute slot, in layman's terms.

Twenty years ago, I had the good fortune to follow a science communication course, where I learned to communicate more effectively with slides, in radio interviews, or on TV. I realized that I actually liked it, and in the intervening years, the lessons learned were of value ... once or twice a year, such as in a podcast as to why the development of medicinal products is so expensive.

That was until Spring 2020 ... Our institute at that point had had a long-standing tradition of work in infectious diseases, with output in the area for nearly three decades. In the most recent half of that period, a mathematical modeling of infectious diseases hub was built by Niel Hens and a large and growing team around him. When SARS-CoV-2 struck, the team was involved from Day 1 in Belgium's pandemic response. Every aspect of our normal academic operation was put under severe stress, and our lives were lived at twice the normal speed. We had to reinforce the research capacity, while organizing working from home and switching over all teaching to an online format. In March 2020, Niel Hens was appointed to a government COVID advisory board. One of the roles I took up was media and press communication. We thought that this would be needed for a few weeks or perhaps months, but it continued for well over two years. In spring and early summer of 2020, I did not have a formal advisory role, so I could concentrate on science communication. This involved explaining the 'corona weather' of the moment, exponential growth, flattening the curve, etc.

Things changed in the summer of 2020. After successfully counteracting the first, severe wave, arguably too enthusiastic a relaxation of non-pharmaceutical interventions, and re-opening of international travel within Europe, numbers rose again. In mid-summer 2020, I was appointed to scientific advisory boards myself, and this has not stopped at the time of writing.

Communication became several magnitudes more complicated. The impact of the pandemic and the unavoidable non-pharmaceutical interventions cannot be underestimated, so every statement should be checked against the impact it will have on the population. There is no benefit to be gained from instilling fear. On the other hand, too euphemistic a communication will fail to be captured. Academic freedom of speech and loyalty to the authorities who establish a scientific advisory board make for a difficult marriage. If two scientists bring the same message in slightly different terms, certain media will frame that as opposing views – a hard lesson to learn.

In Belgium, like just about in any other countries, society as we normally know it, organized within a vast and belabored legal framework, was suspended and replaced by ad hoc rules (stay in place orders, perimeters, quarantine rules, isolation rules, closure of national borders). An interesting example is that of limiting contact behavior.

An intuitively evident principle, rooted in profound mathematical and statistical work, is that contacts should be limited in frequency, and those that take place should be protected (via testing, face masks, and ventilation). But when a large population is involved, principles should be turned into precise guidelines. If the *principle* is to limit contacts to four per week, then having five contacts complies with the idea behind the principle, but if the *rule* is to limit contacts to four, then five contacts mean that the rule is violated.

When restrictions are going to be alleviated, the typical societal machinery clicks into gear whereby all sectors and lobby groups argue that their sector should be in the front row to receive relaxation. One of my assignments was to communicate with the education, higher education, tourism, and air travel sectors. Often, we were not the messengers of good news … The best one can do is to describe the situation, a bit like in a weather forecast, and then point to the risks and benefits of certain strategies.

Lesson learned: While many people may be willing to display common sense and apply a principle, a large population as a whole may not be. The only option in a pandemic is to apply clear and well-communicated rules.
Lesson learned: We bring advice to policy makers, display the options with advantages and disadvantages. It is up to the policy makers to decide. A healthy principle thereby is: 'Comply or explain.' It is not up to us to applaud when they follow our advice or point our thumb down if they do not.

One way to think about a pandemic is as a 'crowd disease.' Like with most diseases, we have to think about what therapy to apply. Therapeutic effects and side effects have to be weighed against each other, like with an individual patient. A strict lockdown, including school and workplace closure, is very effective from an epidemiological point of view, but there are consequences for mental health and the economy. This cannot be done by a single person. Advisory boards, acting like the 'doctor for the crowd,' should be composed of a variety of disciplines – virologists, infectiologists, vaccinologists, immunologists, biostatisticians, epidemiologists, psychologists, sociologists, economists, anthropologists, etc.

I learned that the peacetime organization of our institute, with a lot of responsibility given to all colleagues, shared leadership, and hard work, helped to integrate some of us within the advisory. Both policy work and communication should be solidly rooted in scientific research. Early in the pandemic, we wrote an article on how the statistics community contributed to scientific research and, in reverse, how the pandemic impacted (and simplified) certain procedures (Molenberghs et al., 2020).

When we give seminars or presentations at conferences, they are ideally followed by a good solid discussion. Occasionally, we learn more from the discussion than the talk itself. Scientific debate can be fierce at times but, when kept respectful, that is how progress is made. However, scientists should not do this in exactly the same way in the public forum. It simply does not work. A small difference in opinion will be magnified into a big conflict. This should not be taken to mean that there should no longer be a debate and that 'official' science advice should be followed blindly, including by scientists not on the advisory boards. Critical reflection, peer review if you wish, remains vital also here. But a few principles should be kept in mind.

Speed trumps perfection. When the curve is growing exponentially, it grows fast! There may not be time to assess every aspect of a proposed measure or intervention. After translating all available evidence to the current situation (e.g., masks that have shown their effectiveness toward earlier respiratory infections are likely to be beneficial here as well), and invoking the cautionary principle, the measure can be applied, even though uncertainty remains. As statisticians, we should remind ourselves that we are in the driving seat when it comes to understanding, quantifying, and communicating uncertainty.

Measures imposed should be evaluated incessantly. Perhaps, we overemphasized the importance of hand sanitization. While useful for many purposes, it is not the crucial measure against SARS-CoV-2. It was OK to apply it, but later strong emphasis on this measure became an excuse for relieving other measures. On the other hand, the role of aerosols was underestimated for some time. In the future, the aerosol route should be kept on the drawing board until convincing evidence of its presence or absence is available. For example, when the Province and City of Antwerp imposed a curfew in July 2020, a first in a peacetime western society for over a century, there was no randomized trial evidence that it would work – but it did, according to a post hoc evaluation. A nuance is that it was perhaps applied too broadly, and certain more rural areas of the Province could have been excluded from it.

Lesson learned: Le mieux est l'ennemi du bien (Voltaire).

As was the case with the Spanish flu, the COVID-19 pandemic and its measures were met with increasing skepticism by important parts of public opinion. Scientists on the public forum, especially those serving on advisory boards, have been the target of at times vitriolic campaigns on social media. When I started my career, I did not anticipate that one day I would have an email folder with over a thousand hate messages and that I would be given police protection for about 20 months.

It made me realize that a few boundary conditions had to be met before publicly communicating in the midst of a public health crisis: One needs to like science communication and acquire a certain level of it, but one also needs to be able to deal with the backlash in the form of hate and threat. A collegial atmosphere in the advisory bodies as well as in one's own research environment is a crucial antidote.

Some more reflections on this extraordinary period are given in Molenberghs (2023a, 2023b). The report by the Royal Dutch Academy of Arts and Sciences (2022) is a very important document in this respect.

> Lesson learned: If you like science communication and outreach, go for it. Formal training is useful, but everyone develops their own style. Some people can deal with negative, anti-science reactions better than others. Brief, some people are made to function well in a crisis, whereas others shine in more quiet times.

14.8 INTERACTING WITH COLLEAGUES IN A DIFFERENT SETTING

As applied statisticians, we tend to meet and work with colleagues from a variety of fields, at a rate above average for science disciplines on the whole. At the same time, I really enjoy interacting with colleagues in roles that have less to do with statistics. For example, taking a communication course led to long-lasting friendships with several other participants at the time.

On a more structural level, I have been serving as chair of the UHasselt's Committee on Scientific Integrity for three consecutive terms. Meetings and conversations are necessarily very delicate and confidential, require empathy for the people involved, and require a functional understanding of the customs and traditions in a variety of fields, whether in science or technology, biomedical sciences, or humanities and social sciences.

When taking up this role, it helped that I had served for about a decade on an interdisciplinary panel of the Belgian Science Foundation, which had made me appreciate the very different publication cultures across fields. For example, in mathematics, alphabetically ordering the authors is common practice, unlike the biomedical tradition with specific value attributed to the first and last authors. Many computer scientists publish extensively in peer-reviewed proceedings, which is equivalent to peer-reviewed journals in other fields.

I have thoroughly enjoyed the ability to contribute to the mentoring and coaching programs of my home universities. Often, the mentee comes from a different faculty or institute, which ensures that one can focus on the mentor–mentee relationship, not convoluted by the day-to-day issues in one's own neck of the wood. Learned societies, such as the International Biometric Society and the American Statistical Association, to name a few, have also set up mentoring programs, allowing mentor–mentee pairs to come from different institutions, often even from different countries.

> Lesson learned: Professional interaction does not need to be restricted to statistics only.

And then, it is always joyful to meet colleagues that share the same hobby or pastime. Let me conclude with a few words about hobbies.

14.9 HOBBIES

Some people are so passionate about statistics that it fills pretty much their entire active life. That's perfectly fine, but for me personally having several hobbies has always been the icing on the cake.

Books have continually played an important role for me, fiction and non-fiction. From primary school onwards, I enjoyed reading – and re-reading – fiction. From early on, I grew a list of favorite authors. To honor the work of one of them, Hubert Lampo, who wrote in Dutch but with much of his work translated, there is a literary society of which I am a member.

Books are written and for that we use language, another hobby of mine, especially the Dutch and English languages. I am one of those people who can enjoy reading a text on English grammar or browsing through a dictionary. Being a native Dutch speaker, English is a foreign language, but with sufficient grammatical proximity to Dutch so as to develop a good feel for it. Also, I have been fascinated by its complex history, so the grammatical texts are supplemented by texts on the language's history and evolution.

This brings me to history, which in my opinion in a tangible way adds a fourth dimension – time – to our three-dimensional world. I am fascinated by the 16th century in my native city of Antwerp, its gilded age, like the 17th century was in the northern part of the Low Countries, in particular Amsterdam. This brings me seamlessly to New Amsterdam and New Netherlands, the Dutch roots of New York City, and the surrounding area. I decided to become a member of the New Netherland Institute.

New York City has become a hobby in its own right. The three geographical dimensions of the city are plenty as such – I am not only referring to the skyscrapers, but above all to the bedrock (Manhattan schist)-shaped terrain of Upper Manhattan and the Bronx. The fast-paced history of the city over the last 300 years, with an incredible growth rate in the 1800s, never ceases to captivate the interested amateur that I am. To top it off, New York is a Walhalla for a public transportation – mass transit – enthusiasts. The interest in public transport, especially urban rail transport (tramways, underground systems), started when I was

FIGURE 14.1 Spuyten Duyvil railway station on Metro North, Riverdale, the Bronx (left; September 15, 2018)), the Palisades, viewed from Washington Heights (middle; October 22, 2022) and Highbridge, Harlem River Drive, Manhattan (right; September 15, 2018).

about seven and has never left me. There is of course something very geometric about a time-honored extensive rapid transit rail system, such as London's Underground (Tube), the Paris Métro, and Gotham's subway. At this point, a large number of cities and regions have such a system, but the complexity of New York's system, organically grown out of the merger of three distinct systems, is an endless source for exploration. The reader might not be surprised that I am a member of the New York Transit Museum.

The interest in cities blends in nicely with my regular habit of running. The joys and advantages of running for physical but also mental health are well known. In addition, it allows one to explore a city (and beyond), fast enough to get to see large portions, and slow enough to take in what you see. In New York, this means running out and riding in, while looking around and thinking, the latter sometimes but now always about work (Figure 14.1).

CHAPTER 15

A Final Word

ABSTRACT

Other chapters refrained from specific advice due to the unique and varied characteristics of careers in statistics. This chapter deviates from this general rule and provides a brief summary that includes specific actions step for those who wish to pursue steps to help build a long, happy, and accomplished career. Although there is no single plan for success in statistical careers, the points outlined in this chapter will be a good place for many to begin.

Christy, Marc, and Geert's careers, and mine, were very different, yet some important things were common to all. Each career shows:

- Highly motivated
- Means used to be productive over prolonged periods
- Collaboration and exposure to diverse ideas
- Emphasis on communication
- Effective work relationships

DOI: 10.1201/9781003334286-20

- Being coached and mentored, providing coaching and mentoring

- Continued learning

Knowing what others have done is useful, but what exactly are YOU supposed to do? We refrained from specific advice throughout this book because careers are so diverse and idiosyncratic. But I think it is worthwhile to suggest a few starting points that would be useful for almost everyone.

- To assess where you are in your career development today, ask these three questions. (1) Am I motivated in my work? (2) Am I working smart (using attention management and other approaches for productivity and creativity in Chapters 2 and 3)? (3) Am I working well with others? If the answers to all three questions are yes, you are on the right path. If one or more answers is no, consider making the necessary changes. If you do not know the answers to one or more of the questions, seek feedback.

- When considering where you want to go in your career and how you are going to get there, consider the song lyrics below.

It's a good question to be asking yourself
What is the good life, what is wealth?
What is the future I'm trying to see?
What does that future need from me?

– JACKSON BROWNE

What is the good life, what is wealth? This is how you want to live and what things you value. What is the future I'm trying to see? This is your grand purpose goal(s). What does that future need from me? These are the things you can do to help achieve the grand purpose and the specific things you want to achieve.

Don't be surprised if things don't go as you had planned. They never do. Remember, it's a journey, not a destination.

References

Alonso, A., Bigirumurame, T., Burzykowski, T., Buyse, M., Molenberghs, G., Muchene, L., Perualila, N. J., Shkedy, Z., & Van der Elst, W. (2017). *Applied Surrogate Endpoint Evaluation with SAS and R*. Boca Raton, FL: Chapman & Hall/CRC.

Anderson, C. (2016). *TED Talks: The Official Ted Guide to Public Speaking*. Boston, MA: Houghton Mifflin Harcourt.

Bailey, C. (2018). *Hyperfocus: How to Be More Productive in a World of Distraction*. Toronto, ON: Random House Canada.

Brecht, B. (1986). *The Life of Galileo*. London: Bloomsbury Publishing PLC.

Brown, K. G. (2013). *Influence: Mastering Life's Most Powerful Skill*. Chantilly, VA: The Teaching Company.

Burzykowski, T., Molenberghs, G., & Buyse, M. (2005). *The Evaluation of Surrogate Endpoints*. New York: Springer.

Carstensen, L. L., Pasupathi, M., Mayr, U., & Nesselroade, J. R. (2000). Emotional Experience in Everyday Life Across the Adult Life Span. *Journal of Personality and Social Psychology, 79*(4), 644–655.

Castelvecchi, D. (2015, May 15). Physics Paper Sets Record with More Than 5,000 Authors. *Nature*.

Chuang-Stein, C. (1996). On-the-Job Training of Pharmaceutical Statisticians. *Drug Information Journal, 30*, 351–357.

Chuang-Stein, C. (2017). Improving Our Communication and Presentation Skills. *Biopharmaceutical Report, 24*(2), 1–4.

Chuang-Stein, C. (2019). Be Bold. *Biopharm Report, 26*(1), 14.

Cialdini, R. B. (2008). *Influence* (5th ed.). Upper Saddle River, NJ: Pearson.

Collins, J. (2001). *Good to Great*. London: Random House Business Books.

Collins, R., Bowman, L., Landray, M., & Peto, R. (2020). The Magic of Randomization and the Myth of Real-World Evidence. *New England Journal of Medicine, 382*(7), 674–678.

Dmitrienko, A., Molenberghs, G., Christy Chuang-Stein, J. L., & Offen, W. W. (2005). *Analysis of Clinical Trial Data Using SAS: A Practical Guide.* Cary, NC: Sas Press.

Doyle, G. (2020). *Untamed.* New York: The Dial Press.

Duez, N., Eeckhoudt, D., Kirkpatrick, A., et al. (2013). Professor Maurice Staquet, MD, MS (1930–2013). *Statistics in Medicine, 32,* 5219–5220.

Ellenberg, J. (2014). *How Not to Be Wrong: The Power of Mathematical Thinking.* New York: The Penguin Press.

Ellenberg, S. S., Fleming, T. R., & DeMets, D. L. (2019). *Data Monitoring Committees in Clinical Trials: A Practical Perspective.* New York: John Wiley & Sons.

Fisher, J., & Nguyen-Phillips, A. (2021). *Work Better Together: How to Cultivate Strong Relationships to Maximize Well-Being and Boost Bottom Lines* (1st ed.). New York: McGraw Hill.

Goodwin, D. K. (2006). *Team of Rivals: The Political Genius of Abraham Lincoln.* New York: Simon & Schuster.

Goodwin, D. K. (2018). *Leadership: Lessons from the Presidents for Turbulent Times.* London: Viking.

Grant, A. (2016). *Originals: How Non-Conformists Move the World.* London: Penguin Random House.

Howard, R. A. (1966). Decision Analysis: Applied Decision Theory. In D. B. Hertz & J. Melese (Eds.), *The Proceedings of the Fourth International Conference on Operational Research.* New York: John Wiley & Sons, Inc.

Johnson, S. (2011). *Where Good Ideas Come from: The Natural History of Innovation* (1st Riverhead trade pbk. ed.). New York: Riverhead Books.

Kahneman, D. (2011). *Thinking, Fast and Slow.* New York: Farrar, Straus, and Giroux.

Keller, G., & Papasan, J. (2012). *The One Thing: The Surprisingly Simple Truth Behind Extraordinary Results.* Austin, TX: Bard Press.

Lencioni, P. M. (2002). *The Five Dysfunctions of a Team.* Hoboken, NJ. Jossey-Bass.

Mallinckrodt, C. H. (2019). Craig's List. *Biopharmaceutical Report, 4,* 3–7.

Mallinckrodt, C. H. (2020). Craig's List. *Biopharmaceutical Report, 27*(1).

Meunier, F., & van Oosterom, A. (2001). Obituary: Professor Henri Tagnon. *European Journal of Cancer, 37,* 551–552.

Molenberghs, G. (2005). Presidential Address: XXII International Biometric Conference, Cairns, Australia, July 2004: Biometry, Biometrics, Biostatistics, Bioinformatics,... Bio-X. *Biometrics, 61,* 1–9.

Molenberghs, G. (2023a). The Role of Biostatistics in the Response to COVID-19: A Belgian and International Perspective. *Israel Journal of Health Policy Research,* to appear.

Molenberghs, G. (2023b). Biostatistics and the COVID-19 Pandemic in Belgium, in 2020 and 2021. *Statistique et Société,* to appear.

Molenberghs,G., Buyse, M., Abrams, S., Hens, N., Beutels, P., Faes, C., Verbeke, G., Van Damme, P., Goossens, H., Neyens, T., Sereina Herzog, Theeten, H., Pepermans, K., Alonso Abad, A., Van Keilegom, I., Speybroek, N., Legrand, C., De Buyser, S., & Hulstaert, F. (2020). Infectious diseases Epidemiology, Quantitative Methodology, and Clinical Research in the Midst of the COVID-19 Pandemic: A Belgian Perspective. *Controlled Clinical Trials, 99,* 106189.

Molenberghs, G., Thijs, H., Jansen, I., Beunckens, C., Kenward, M. G., Mallinckrodt, C., & Carroll, R. J. (2004). Analyzing Incomplete Longitudinal Clinical Trial Data. *Biostatistics, 5,* 445–464.

Molenberghs, G. & Verbeke, G. (2005). *Models for Discrete Longitudinal Data.* New York: Springer.

Newport, C. (2016). *Deep Work: Rules for Focused Success in a Distracted World* (1st ed.). New York: Grand Central Publishing.

Porter, T. M. (1986). *The Rise of Statistical Thinking 1820–1900.* Princeton, NJ: Princeton University Press.

Rao, C. R. (1989). *Statistics and Truth: Putting Chance to Work.* Fairland, MD: International Co-Operative Publishing House.

Roberto, M. A. (2013). *The Art of Critical Decision Making.* Chantilly, VA: The Teaching Company.

Royal Netherlands Academy of Arts and Sciences (KNAW). (2022). *The Pandemic Academic: How COVID-19 Has Impacted the Research Community.* Amsterdam, the Netherlands.

Ruiz, D. M. (1997). *The Four Agreements: A Practical Guide to Personal Freedom.* San Rafael, CA: Amber-Allen Publishing.

Spetzler, C., Winter, H., & Meyer, J. (2016). *Decision Quality: Value Creation from Better Business Decisions.* Hoboken, NJ: Wiley.

Strunk, W., & White, E. B. (1999 [1959]). *The Elements of Style* (4th ed.). Boston, MA: Allyn & Bacon.

The GUSTO Investigators. (1993). An International Randomized Trial Comparing Four Thrombolytic Strategies for Acute Myocardial Infarction. *New England Journal of Medicine, 329*(10), 673–682.

Twain, M. (1906). *Chapters from My Autobiography.* Oxford: Benediction Classics.

University of Minnesota. (2010). *Organization Behavior, Module 13.2, the Basics of Power; Module 13.3, the Power to Influence.* The Power of Influence. Accessed March 9, 2022. https://open.lib.umn.edu/organizationalbehavior/chapter/13-3-the-power-to-influence/

Venet, D., Doffagne, E., Burzykowski, T., Beckers, F., Tellier, Y., Genevois-Marlin, E., Becker, U., Bee, V., Wilson, V., Legrand, C., & Buyse, M. (2012). A Statistical Approach to Central Monitoring of Data Quality in Clinical Trials. *Clinical Trials, 9*(6), 705–713.

Verbeke, G., & Molenberghs, G. (2000). *Linear Mixed Models for Longitudinal Data.* New York: Springer.

Index

Printed in the United States
by Baker & Taylor Publisher Services